CONSTRUCTION

SARIPALLI SURYANARAYANA, B.E., FIE.FIV.
LM-ICI., Fellow-ACCE[I], IIBE, Sr.P.E.[ECI]

sincere thanks – And   Acknowledgements To
To   My wife   Janaki, sons Dr.Srikanth and Mr.Sricharan,

Contact for copies.
Saripalli suryanarayana
D-1,CHAITANYA APARTMENTS
RAJENDRANAGAR,VISAKHAPATNAM-530016,A.P.,INDIA
Mobile-+919392387088
E-mail,s_n_surya@yahoo.com,
suryasn.saripalli@gmail.com,

ENGINEERING COUNCIL OF INDIA.,New Delhi

[Partner: United Nations Institutional Outreach Platform]

ECI was established in April 2002 as the apex Body of engineering profession in India by coming together of a large number of Professional Organizations / institutions of Engineers to work for the advancement of engineering profession in various disciplines and for enhancing the image of engineers, in society, by focusing on quality and accountability of engineers. Today there are 33 Indian Engineering Professional Associations/ Institutions as members, representing practically all major engineering streams and with 8 associated government departments' nominees on its Board.

In order to improve the employability of the students by imparting required skills and making them industry ready, it has been decided by AICTE to introduce mandatory internship for students from the session 2017 – 18. **MoU between AICTE and ECI**

Dr.P.R.Swarup-Member secretary-Engineering Council of India, and Director General,[CIDC] Construction Development Industry Council. Dr.P.R.Swarup is extensively connected with the construction Industry during last 5 decades.

www.rkecprojects.com , **RKEC Projects Ltd** is a Construction Company, over 37 years old, specializing in civil and defense construction, such as the construction of Buildings, Highways, Marine Works, and bridges. During the last decade, the company has completed many construction projects in India covering Andhra Pradesh, Tamil Nadu, Gujarat, Odisha, Rajasthan, Maharashtra, Puducherry, Kerala, Manipur, and Uttar Pradesh. In addition, the company has expanded our presence to the state of, West Bengal, Jharkhand, and Karnataka with ongoing projects. **Chairman and Managing Director**

Shri Garapati Radhakrishna.

ASSOCIATION OF CONSULTING CIVIL ENGINEERS (INDIA)

Located in Bangalore.To foster ideals of profession, promote friendship, establish rules for professional and ethical conduct and to develop social awareness and responsibility in the members. To encourage and foster the ideals of the profession. To hold conferences/meetings/seminars for dissemination of knowledge among the Civil Engineers in particulars and society in general.

Communities of Practice;-The next generation of UNDP's Communities of Practice (CoPs) is a distributed network of thematic experts and practitioners who collaborate to define, recognize, and solve specific development challenges.The new CoPs are structured along with UNDP's Signature Solutions with the aim to connect knowledge workers through moderated discussions.

1.CoP on Keeping People out of Poverty and Reducing Inequality.2.CoP on Governance for Peaceful,just and inclusive Societies.3.CoP on Crisis Prevention and Resilience.4.CoP on Nature Based solutions and Climate Action.5.CoP on Clean, affordable Energy.6.CoP on Gender Equality and the Empowerment of Women.

SparkBlue is curated by UNDP's SDG Integration team

Dr.Jan Goossenaerts.Phd.speaker on 12 th august 2023 the Topic;- " An introduction to societal architecture and its promises". He has established the Wikiworx Academy · a philosopher for Societal architecture, PhD Mathematics, MSc Informatics, Certificate Philosophy. Consultancy and facilitation

on requirements-led governance and architecture of public-private service solutions, using instruments such as skill frameworks, action research, regulative cycle, enterprise architecture, decision frames in multi-stakeholder settings. Recently launched the wikiworx platform, a platform for the catalysis and carrying of systematized content commons for inclusive and sustainable development.Degrees in mathematics (B. Sc. (1982) and Ph.D. (1991]

"Deception of workers into Labor Trafficking in Global Engineering Projects" was the speaker on 30 th July 2023,on a webinar. Kalyani Gopal, She is Author, Clinician, Keynote, CSA, Human Trafficking. President, Society for Clinical Psychology, APA. Pres. @SAFECHR , Top 20 Global Women of Excellence Medical & Health Munster, From ;Indiana, US.[SAFECHR – Human Trafficking Awareness, Freedom and Empower… https://safechr.org This is their organisation.

[SAFE Coalition for Human Rights] (SAFECHR)This is An Accredited NGO with Special Status with the [Economic and Social Council] ECOSOC of the UNITED NATIONS.

Expert and affected by the Traffickers is Harold D'Souza, President, Eyes Open International- Survivor of Labor Trafficking-Also has spoken on the occasion.[On July 30th on the U.N,Anti HumanTrafficing Day]

3

A jetty which was constructed and handed over

Jetty under use after construction

Not a Preface but Introduction

The events and time line [dimension less] in this book are actual. All other things are fictitious. Acknowledge the sensitive ness of the events, and with due respect all facts are acknowledged. Except for events nothing belongs to me as much unethical use or Plagiarism are avoided. Along with all writers mentioned I acknowledge John Grisham, if I had copied his style anywhere. Creative nonfiction or "Gonzo journalism" is a journalism that is written without claims of objectivity, often including the reporter as a first-person narrator.

'For a text to be considered creative nonfiction, it must be factually accurate, and written with attention to literary style and technique. "Ultimately, the primary goal of the creative nonfiction writer is to communicate information, just like a reporter, but to shape it and it's attitudes correctly"
The word "gonzo" is believed to be first used in 1970 to describe an article by Hunter S. Thompson, who later popularized the style. Gonzo journalism tends to utilize personal experiences and emotions to achieve an accurate representation of a phenomenon. "http://en.wikipedia.org/wiki/Gonzo journalism, Posted by Jonathan Post.
I am using a language which I foresee as the futuristic, I am using futuristic models that fits in to Engineering, if not in to science. After all engineering changed lives in last 300 years, and last known 55 years we have seen huge advances in mobility and use of technologies for betterment.

'This book is dedicated to aspiring Engineers and Project Managers world over. These are people working under tremendous social tensions, leaving their families, and living for a single cause that is completion of a project. Travel and communication have vastly improved, from 1968 to today.'

Er.Saripalli Suryanarayana.[SURYA. S.N]

Again'This book is dedicated to project Engineers and workers world over. These are people working under tremendous social tensions, leaving their families, and living for a single cause, that is completion of a project. Travel and communication have vastly improved, today.'

CONSTRUCTION

CHAPTER ZERO

Construction industry requirements and training man power. Are present and futuristic. Given the last 50 years transformation we have to prepare for the changes. The earlier civilizations have adopted the water supply and sanitation as prime objects of living. Today industrialization, commutation, and scrapper buildings are leaving us with high density populations ground.

'The weathering phenomena is same for rocks, on which the foundations are founded.

The soil pores, internal canals, and stress remains same.

The igneous and metamorphic rocks offer their systems of changes. The over stressed soils liberate themselves, from the stress. Nature is prenominal. Earth yields, call it a quake or a temporary yield. The stress has to be limited keeping in view next 200 years in view'.

The **trainer** needs to understand the **organizational requirements** in training supervisors, executives, and the workers ultimately. The formal education system cannot visualize where the trained manpower will be placed after their initial education. So also, the vocational education also cannot specify the needs. The needs of any industry are time specific. Technology is the root of all changes in human lives, and living.

Who will train, and who are to be trained. These are fundamental questions we will answer in this book. A step-by-step process of identification of the requirements to train each category; say Barbenders, Carpenters, Store keepers, Accountants, Supervisors, and Foreman are listed in a phased manner. Identifying the needs, and dealing with human psychology in overcoming the stress are vital for highly qualified technicians in off shore jobs, such as welders,

fitters, grinders working in laying and, testing of pipeline for oil and gas. While punishment is ultimate, recognizable gifts, occasionally, along with some awards needs to be given by the organization, as well by the project owners.

UNESCO has asked Draft.3.0 comments and guidelines from individuals, societies and the Governments. It is in relation to the media out lets, which are more powerful, appear to be small but used in excess than the needs of the people for the purpose of knowledge decimation, transfer, safety and whatever possible in the community and governance systems. The travel of news is so fast that Media Information Literacy of UNESCO-MIL, has to step in to contain the fake news travel, and find solutions for the act of governance systems intime to avoid massive side effects in the society.

Internet 4 Trust was latest discussions initiated by the UNESCO and went on in Till almost July 10 th.The final draft guidelines are encrypted. Platform use and spread of lies are centric to find ways and means in society.

SDG 7 and interlinkages with other SDGs – Affordable and clean energy; Energy lies at the heart of both the 2030 Agenda for Sustainable Development and the Paris Agreement on Climate Change. Ensuring access to affordable, reliable, sustainable and modern energy for all will open a new world of opportunities for millions of people through new economic opportunities and jobs, empowered women, children and youth, better education and health, more sustainable, equitable and inclusive communities, and greater protection from, and resilience to, climate change.

Progress on household drinking water, sanitation and hygiene 2000-2022: Special focus on gender has been released. It provides the first in-depth analysis of gender inequalities in WASH. It compiles data on global progress towards achieving universal access to safe drinking water, sanitation and hygiene (WASH), including emerging data on menstrual health and hygiene. For the first time, the report provides an in-depth analysis of gender inequalities, highlighting the risks women and girls face from inadequate access to safe WASH in those countries for which national statistics are available.

More issues related to women wash areas in schools, colleges, including providing safe hygine pads, and water needs during menstrual cycle and premenopausal times needs to be attended.

My novels listed below deals more with modern day hospital constructions for better living.

CHAPTER ONE
MIXING or WASHING

Training the trainer is guided in many manuals, books, seminars. However **training the trainer** for the **unorganized sector, like the construction industry, is still at gross routes.** The need is to understand the changes is imminent.

The other day I had put two brand new blankets in the washing machine. I found the tub is still not full, so I took two used but, very white Turkish towels, along with the blankets. The current or the electricity in between went away. When the [Power] current came back, the washing machine performed excess washing. When the spin was over, while drying the cloth, I found the white towels, turned grey. Nothing changed for the blankets, they were washed well.

The addition of new color is known to me because I had seen it as white, for any other it looks as though the original co lour it self is grey. There are at least two principles in this, any **changes** in the **trained** can be grasped by the organization. There are people who will say we want the sheen to be washed off, and that is why mixing is better. But 'are they sure that the sheen from the brand new one is to be washed off', or ' do they want the old one to look like the new one'. There lies the perception, and then the performance.

When you find a chamber or a work place is not full, donot send different people for this work, who do not mix in to it, the result could be devasting. The key to success is in recognizing the fact that some people tend to lose their originality, while others may attain it for their betterment. **No training is perfect. No theory is ultimate. If it is Improved each time, it could be innovation.**

Dependence, on firm supply is imminent and the market orientation has to be adopted.

I was in a restaurant last month; it gave me three Indian chutneys. All were different. Ingredients were same. The smell and taste differed.

One had the tamarind base
Second has the Lime base
The third has garlic, and ginger base.
And the seller might have more ways of making chutneys. I took only three.

India and Hyderabad do have many festivals apart from many market bandhs, happening for whatever reason. The normal item in such days is the vegetable markets are closed. And vegetables are not available and sometimes next day also. Luckily on many such holidays. We could fetch the vegetables. Thanks to **Reliance fresh**, and **Food world**, even while paying the cost we could get my choice. Hence in market-oriented economy, and time as a factor, we need keep our supply lines open. Do not fore- close the costs? Sometimes we need to pay more to keep our selves moving.

The big might be dependable. We were wondering in one of the **reliance fresh** shops, how they are able to maintain the **logistics** and **supply chain management.** Of course nothing is simple but nothing is difficult, as [Copyright © 2007] *David Weinberger says in one of the chapters in 'Everything is miscellaneous'.* Everything Has Its Places, says Weinberger.

'It won't be easy. The world started out miscellaneous but it didn't stay that way, because we work so damn hard at straightening it up. Take eating, the most basic bodily activity we do on purpose'. Not only that we organize the eating table, we organize the kitchen, and all that matters. How ever if we are in the middle of work, we do not mind having a buffet. Having separate cups for coffee, beer jars needs to be separate with glass. The table need to accommodate the persons intended to accommodate.

When we are **in to construction we need to organize, our logistics,** our **supply chain management**, as we do for our lunch, in our daily life, or as we stack materials item wise in a super store.

In Ethiopia, when in Deddar, a hilly remote, but prominent area. Daily morning the students were coming to class at 8 am, with their tooth brush [neem stick] in their mouth. In few days after careful observation, could make out that the staff and the principal also come with the brush in their mouth.

Well, that was how the world was, disorderly, or mislleneous, we are trying to straighten it. So the students, were told that the college is not for face wash and brushing, and that they need to complete it before they leave their house in the morning. Slowly but steadily in a month it stopped. It is true that our ancestors were not using the modern brushes, for teeth cleaning; it is also true that the best animal cannot clean its teeth, so we have organized. The key to human success is organizing themselves.

When we wanted to buy some window and door curtains for my house, we went to as many cloth stores as possible, Every time we went to the Kirana stores we could not buy all items we wanted. Sometimes, cool drinks, sometimes soaps, and some times, other grocessories were forgotten. But when ever we landed in the departmental stores of some standard, we was coming out with 95% of my requirements. And so was our visit to the cloth stores resembled that of the Kirana stores. Then we decided to see the exclusive Drapes, or curtain stores. The result was sound. We not only got what we wanted but also, found some interesting new varieties. So our marketing was successful.

In a similar event, in 1995 for a project in Kakinada, Beach, we were looking for aggregate or crushed stone, for our concrete work of Tetrapods.These tetra-pods were to be cast at site, and then to be shifted for shore protection along the beach. Each tetra pod was weighting about 10 tons. The concrete mix was to be M30, Sulphate resistant cement was to be used, as the sulphates and chlorides were beyond the acceptable limits in the sea water. Hence, we were searching for granite or basalt stone which are available at Rajahmundry, which was about 100 KM, apart.

The stones were good the quality of crushing was bad because of two reasons. One was the crusher teethes were old and not crushing uniformly. Second the screens were badly installed, and were old. After roaming about a dozen crushers we ultimately could get the approval of Mr.Ayyangar [The consultant] and what is the need for the Kakinada port trust project.

Having done that, we went on arriving at test specimens and tested them as per then prevalent IS 456.The cement used was minimum 400 kg, per cum of concrete. In between we used ACI also to compare. We got the results. Nehru and Mr. Subba Rao from ESSAR were also one way or other involved in quality control of the project.

Consistency in approach is required, even when new technologies are to be adopted

The guidelines have been set by American Society for Training and Development, or ASTD, and Indian society for Training and Development. In his book Robert.L.Craig, which was Published in 1996, by McGraw-Hill Professional good examples and need for training are listed. And such of these researches which can be optimized to suit the conditions for India and other countries are a good knowledge spreading books. [American Society for Training and Development, 2008 - Employee handbooks - 280 pages] [these are another set which demonstrate the needs of the present society]

Train the Trainer is a four-volume collection, containing the best and most popular issues about the training process--from instructional design to ethics to evaluation. Train the Trainer volume 3 provides you with a selection of training programs to implement. The contents of this volume will get you started with facilitation and workshop skills. This volume includes the following 15 issues--How to Facilitate, Effective Classroom Training Techniques,

In all,Mr.Craig,and others have described the involuntary benchmarking process, and its organizational structure. The books have quote from the research done to establish the requirement of 8% training period to the executive, to the professional at 17%, and supervisor 22%, and non-supervisors at 23%.

These statistics might have reduced by 5 to 10 % by now in 2007, and may go-up again to same standard in 2009.These are time specific and project specific for the construction industry.

Reasoning and instinct are necessary tools for a construction man. Just as by Thomas L.friedman, describes in the book 'The world is flat'. This is a book which describes, how the world is taken over by the internet revolution, and contributors for the same are from many nationalities, and majority of them are the Indians. A day cannot pass in these times, without the use of the computers and the internet. The book deals in the technological development of internet, from infranet, Binary numbers, coding and many applications we have emerged user friendly technologies. We have gone ahead and made the android systems, and 5 G technologies and the development of 3 rd stages of computers.

Well, the construction industry is very ancient, and has taken turns and twists. It is the main focus of development. No one wrote how new trends in construction are being taken over by new addition of technology. This is still not invention. It is improvement over the other technology in use. **Use Common sense, and visualize what is happening in front of you.**

My friend's dog was most human friendly. One day we found the dog was moving erratically, and it was tearing clothes, and biting everything available in reach. It was in fact trying to bite the humans also. We said take her to her doctor, she must be having dental problem. My friend himself had the teething problem, and delayed by two days. Third day he found it had bitten his own daughter. He took the dog to the doctor, who treated her for the decayed teeth. So dogs also have similar problems as humans, it is common sense.

In Construction we have theories developed over years, which were tested on ground. The flow through channels, developed by Lacey, and Kennedy at Punjab, and theories by Dr.Khosla, and Dr.K.L.Rao were actually tested on ground for fluid mechanics and irrigation.

The first it was the Mahanjadaro, and Harappa, followed by Greeks, then Romans, followed by many civilizations where the water supply and drainage were developed and tested. However, layout of roads, water drainage, and storage might have started quite some time back. The Archimedes principle, Euler's theories have made the so called shuttering possible for the liquid concrete. There was more common sense in each discovery, followed by theoretical analysis.

Training the owner's and consultant's representatives and engineers, is immanent, to achieve the desired results.

People executing the projects are not the conceptualizers of the project, unfortunately, the success or failure of such a project is reflected on their future. Remember the contractor who did Suez Canal earned good profit and went for doing the Panama Canal.

He became [including his company] bankrupt, due to
[a] unforeseen weather and diseases,
[b]The project required more engineering, and extra 'water locks' were needed to be created, to meet the terrain, and water conditions. One of the chapters is well dedicated to the stress and strain a project creates on the personnel, during its implementation. The key to success is proper engineering, for any project.

There is massive infrastructure development, and urbanization requirement for the future generations. Remember we are organizing from nothing as we proceed forward. As we get urbanized, we also use more machines, and so less labor.

New technologies are constantly developed and are in usage.The person who uses successfully the new technologies is the successful Engineer and the contractor.

In 1970 and in 80's we were seeing the use of blue prints, on

which the Architects and engineers' drawings were available for site execution. The architects and engineers used to present, perspective plans as to how the areas look after a project is completed.

From 2000 and onwards massive use of computers and STUD, Pro-e, and CAD, have fastened the drawing making process. The internet, logging mails have made speed possible in communications and approvals. To day seeing how Ameerpet area of Hyderabad looks, in a drawing will be a laughing stock, as computers and photography are well available.
Taking short photos of work and inscribing them in the mails [WhatsApp] has become a good philosophy. Other tools are making strides in to the 21 st, century, and companies' needs to get fast in to them. The realities lies beyond CAD, CAM, please go to see for yourself the development in the technologies. http://www.drupal.org.is another online learning and discussion form, used by many communities. Interaction between the students and the teacher, and as well the participants and their facilitator, sharing of ideas and asking question, are the requirements of the future. These are e-learning programmes which allows students/ or participants to learn on their own at a convenient time and allows facilitators to evaluate the outcome.

Web 3.0, interactive technologies have been developed, and were in used for better education and social development. **As a matured civil society in 21 st century we all needs to be interactive and shall run the governance' Hence society at large shall be responsible for all changes in life style occurring, now and beyond.**
Web.3.0Technologies are deep digitally, and there is boundary layer surrounding them.

'Enterprise 3.0 , or societal is an important concept that it represents the most important and potentially disruptive business challenge since the advent of modern management.
For better or worse, Enterprise 3.0 has become COMMUNITIES OF PRACTICE [Modern Constructions and Developmental needs]. **Enterprise 3.0** is about semantic social business: the efficient creation, share and access

Now we are in to web 3.0 technologies. Where all our friends are working for the conceptualization of avoiding the digital divide. Better development for the last mile stone, for the end user, for the last man. Digitalized world needs better aware Engineer for latest technologies with 5G, instruments. Needs of Excel, Auto cad and Other modified forms of project management tools are needed. Artificial intelligence or A.I, may be near but far away. So also, Autonomous technologies in pre-casting and hoisting of slipform or other techniques for huge tall housing are not away.

The need for societal architecture and Corroborative decision making as the need for development is best described by the author mentioned below.

Jan BM Goossenaerts, is a writer with the basics of human development. He has qualifications, PhD Mathematics, MSc Informatics, Certificate Philosophy

He wrote an important -book "Societal Architecture and Collaborative Decision Making"; Towards a talent explosion for sustainable development.

A consultant from various fields to the United Nations Development program, he has written about Viki, tag coding and use of Tax systems for better governance etc.

The book has focus on the digital enablers, encompassing, Leadership, Decision making and communication.Open source modelling tools covering various dimensions of public and private decision making.

Excavator in Action, A Crane in background

CHAPTER TWO

Conceptualize what you have seen

I lived in my grand father's own house in my younger days. The house was built with mud walls. Even the foundations were 5 to 7 feet deep at certain locations. The structure was tiled roof house. The ridge of roof was as high as 16 foot. Consider the present standard of providing 10 to 12 feet ridge height; it was really taking care of summer heat. The roof was made of truss, purlins, rafters on top of which tiles were laid carefully, and in the end eve's board protects the roof materials.

When I was kid, I know that my house contained several musical instruments, such as a harmonium, thabala, a violin, a guitar, and hand-held instruments and flute etc, Also metallic hand drums, and other present day systems still in use were used by my grandfather as good as around 80 years back or earlier. I saw them even in 1965 and was trying them.

[Slipform for House construction, safety net around. Ivan technology in use].

I still remember the HMV gramophone melodiously uttering a good old Telugu song ['Eruvaka sagaro][The tilling of land shall continue with the double bullocks tied on a wooden piece apart, and moving a spike to go deep in to the land].There were many Astro palmistry books, along with one or two books written on leaves, in the old house.

The point was when I joined in the institute for civil engineering in 1965 my mother told me my father also worked in the Zamindari, in public works, before becoming a clerk in simhacahala Devasthanam.

Well, our old houses were constructed with mud walls. Plastered with lime and sand mixture.
The roof was rafter and parlin. A truss also was used. The tiles were laid and renewed every 5 years. A eves board was used, and sometimes joining roofs we had a valley and gutter for drainage. The houses were sold in seventies, and fully modernized in say 2000,with the available technologies.
All these were happenings at Vizianagaram, and Visakhapatnam, of north coastal Andhra Pradesh grandfather was a grate devotee of Sriram, and used to sing ['Ramdas bajanalu'] his devotional songs, using all the available musical instruments in the days of 1945-48.

The house contained the books, on family planning, how to avoid fertility and how to use contraceptives. They were published in 1950.
Keep sense of humor and sense of responsibility

Well, my new education gave me a job, which my father has learnt and did about 25 years ago, when no formal engineering education was available.

Every set of modification in life styles in our countries has required great lengths of education. But for two family professions, which are administrative services, and the other, political services, there has been no change in the requirement of the qualifying degree, for both these professions.

We were travelling in railway compartments in 1965.It was Howrah to Madras mail. It has stop at Tuni, and smalkot. My uncle was a head typist in the Divisional manager office. We used to take his radio along with us, so that we can get the entertainment there. Radio itself was a premier entertainment.

Railways found many changes over what we learnt about the track, and gauges. The compartments, reservation systems were modified. No doubt they were multiple innovations, from 1865, where we cannot write or remember each invention the name of the inventor.

My mothers brother was in the Revenue department as revenue inspector. Every year, in may they had a system of updating

ownership, and payment of cess/or the tax has to be on the records of the lands. That revenue in those days was more important for the State.

The endurance has come to meet the demands of more population. Cheaper houses can be built at about Rs. 250 /- per sq.ft, in semi urban areas. [Or at 6 $per sft]. However land availability is a perennial question.

View what you can accomplish collectively with out losing your identity, in the project

Sustainable development is a factor for a society to prosper, our old houses engineered by our fore fathers has seen 80 to 100 years. Their foundations are a testimony for sustainable development. They had required little maintenance over years, might be they have not kept up for modern requirements of bath rooms, but, for ventilation, and air circulation they were the best.

1] In 2006, February at my apartment in Yellareddy guda, Hyderabad, I was looking at concreting of our road. I could see a mixture machine, sand very big size coarse aggregate or metal, very fine river sand, and cement. There were workers, masons, carry trolleys, and plate vibrator.

Two things were missing from the scene. Graded aggregate and good compaction. Consequently the concrete even after good compaction was a loosely laid concrete. The supervisor/engineer, along with blended or graded aggregate could have done lot of difference, to the strength of concrete.

2] In the year 1994-98, I was working for a good big company in Visakhapatnam. Our job was to construct the palletization plant. There was an approach bridge laid in marshy land with bored piles, by the port authority somewhere in 1975-77.We had seen that with many precautions the soils in the marshy areas were behaving irregularly in 1970-77.The port bridge was no different story, as it was in the same soils, and continued swing at two three spans when even a small car was passing. The ultimate from our experience in marshy land piling in 1971-73, &1994-96 is that; first do not accept any strata other than rock as a founding stratum. Second the friction piles tend to fail, as the raise of water level in may-June by 2.0 M, and with drawl of water in dec, by 2-0 M, both induce soil stresses, and N=100, for founding stratum is a false criterion.

The team, where you may work will have some juniors who anticipate rising fast at the cost of others; hence unknowingly they will accept any criteria.

Be sensible in working in your job. My very higher up, seniors like Colenel.Barve and team had it in 1972-73.

We had it in 1996.The soils are same, and the area is same as it is a vast back filled marshy land, only the working organizations were different. Technology improvement over 23 years has not been utilized effectively, to get correct picture.

As humans no one is special and we a group of engineers met our enlightenment, when the bored piles, at the pellet plant Visakhapatnam, started yielding even without loads. Some of us though were only constructors, and not designers have to take the blame. It happens when the mighty king is always right.

In both the above cases the missing link is the consultant. All these three small examples we earnestly try to put the requirement for a right engineer, and an engineering consultant.

In 1997 in Visakhapatnam, people were working on a neighbor's housing complex. Engineers from the architect's office came for checking the shuttering, supports, and for supervising the concreting, they asked first for a water tube, for level checking, along with a rectangle, for checking the corners. Then they said at least get a leveling instrument. As in the first one the response from the contractor was zero. At last , they said please get me a tape and the above instruments to check and proceed with the work. The Engineer could start checking only after 3 days, and allowed the concreting. Many may be having knowledge about the, fallen centering laid for slabs, and might have seen or heard about fallen concrete slabs, but the latest was the community hall at steel plant site at Visakhapatnam, in November 2006.

This is another case where the understanding of construction by the contractor being negligible or zero. In many such cases engineering the support material used may be nil and zero.

I have praposed in my first appearance in 1989 at Hazira,Relience project.Filling by dredging the river as the vessal carrying capacity for incoming raw meterail is needed.That have also made us to think of a diaphram wall to the neighbours plant.It was handed over to the experienced engineer from a new construction company.The result was not up to the expectations.But it served the purpose for

long term. Ways of Training the Engineer and the Project Manager.

"The Big Rocks." Under which the three points are: (1) growth in revenue and profit, (2) customer service, and (3) operational excellence. This is where the construction manager and his engineers need to be trained.

I had it in 1990-92, sleep less nights. Design of staging, form work for a huge steel plant at TISCO.The blast furnace complex, including the hot stoves, chimney,and ladle above floor etc,are typically designed. The steel needed for staging the floor beams at 10 m height and casting a huge slab of more than 150mx30 m size was 440 m tons. Reputed companies said no to cost time and materials indented. Challenge initiated and completed in time. Ahead of schedule for installation for equipment.

[Dumping and Levelling for a working road in River etc].

Everything is inter-linked, the morning followed by night sleep arrives, it could be cloudy, hot and humid, or may be rainy. However, the day follows each night, so also the construction, growth in turnover, growth in profits are followed by trained manpower, good machinery, good materials, along with best methods are part of the manager's, and engineer's work.

Training the trainer and developing human resources in any organization, is for growth. Growth of the individual, growth of the organization in volume, and in profitability, is what is generally perceived. As said earlier these are the big rocks. How ever the small material, which is like fine aggregate in concrete comprises of many site

19

management issues, which are dealt separately. However there is a high strength bonding agent, like cement and a reaction agent like water which are more are less based on human psychology. These are individual growths, education for betterment, in dealing with his clients, and ultimately the retirement benefits, and the family welfare.

Who shall be the trainer-The organization has to choose, depends on its spectrum of work of such a company. Take for instant a company doing only real estate or apartments constructions, it needs minimum equipment. It's requirements for a total station are limited, and hence it needs only an ordinary surveyor. Its needs for centering shuttering, scaffolding are routine and needs to be articulated. The training for workers needs are limited to good carpenters, good masons, and of course good wood workers and painters.

That is why training needs are industry specific and time specific.

Take the instance of a refinery project, gross root level, for refinery expansion the needs in construction are quite different. The need is to have good amount of hoisting cranes, land leveling, land filling and compacting equipment, along with earth work equipment.

Hence the training has to be on use of equipment, in an efficient manner, with minimum excessive work. Turn over dependence on equipment is also very high, for such works. Here the need is to project skilled surveyors, for each phase of daily work, including bolts, alignment, grouting of foundations, and the like. The use of different set of workers, including bar benders is also high. Hence the training for engineer and the worker of all categories has to be of good standard

CHAPTER THREE

True, that everything was miscellaneous as per David Weinberger, the web designer. We have invented slowly to get urbanized living. It is paradoxical that we organize the unorganized. It is a necessity that we conceptualize on creating a learning system, for better organizational management. What it was yesterday does not matter, but how well we go up matters. So creating a learning organization is first imperative thing.

Employee Performance, quality standards, input training, cost control, inventory, and cost recovery. The rope is called organizational management, and ISO certification needs all these. In fact for renewal also all these and associated parameters, along with safety, are vital. So discussion shall be around who shall be the trainer, who will be trained. And what are the training methods,

Needless to say every week on job training of one to 3 hours on safety, and 4 hours on job training for carpenters, bar benders is necessary, at least for first 4 months.
Civilization as it prospers tends to get urbanized. The 21 st century urbanization will be web-3.0 technologies and higher. The Collaborative Technology Requirements for Social Change needs to be assessed in phased manner.

A GLOBAL JAM was sponsored by the following Organizations: Knowledge in the Public Interest, Impact Alliance, One World, Full Circle Associates, and The Learning Catalyst, on. July 18-19 · 2007

What is a JAM? A JAM was an online discussion that is focused on work, time limited, asynchronous, moderated, and subsequently analyzed. In this unknown but intelligent people are allowed to join.
Today it is a webinar, we use,Meta,Teams,Zoom,Webex etc for our meetings to communicate beyond the U tube and other sources and Face book are still popular.

This chapter we will discuss how on line communities can use the collaborative technologies, for betterment.

On December 3, 2005, a global JAM was held on habitat.Mrs. Anna Tibaijuka, Executive Director, UN Habitat was moved, in an exchange in the JAM on the question "why the habitat jam??". JAMMING with participants from around the world, Mrs. Tibaijuka said this about the power and potential significance these events could have in driving decision/making:

"You are quite right that the problem of slums will not be solved by research alone". The fact that "the debate on slums has moved from the academic world to streets of cities such as Nairobi, Dakar, Cape Town and Mumbai, Rio, Lima and Manila is in and of itself a powerful signal to world leaders on the need for concerted action."

The web via a common data structure called the Resource Description Framework (RDF) is now using the meta systems and converting and giving each type of data a unique, predefined, unambiguous tag.[RDF standard syntax allows the software to efficiently. RDF graphs provide much more information use metadata and exchange information more about entity relationships than relational.
[1]You can prefer to Read my Novel on AMAZON, KDP or paper back. Where use of technologies is mentioned. NIRVANA 2020: 20 TH SEPT 2020;By SURYA NARAYANA SARIPALLI

View on Amazon,KDP
ASIN: 1521056757

Rs1000/-

NIRVANA 2020: 20 TH SEPT 2020 https://amzn.eu/d/bVmTQ7W
NIRVANA 2020: 20 TH SEPT 2020 https://amzn.eu/d/bVmTQ7W

No doubt we are using the web asking the Google or the alexa the questions we never anticipated.

Now the Chat GPT or others also came into the field.

[2] Nani: An Engineer https://amzn.eu/d/1wtfO00

[3] 2025-DIAMOND TREASURE ISLANDS: DIAMOND TREASURE ISLANDS OF ANTARCTICA. https://amzn.eu/d/5PbF1tz

Are presently on AMAZON KDP

"It means that you will be able to ask a website question you couldn't ask before, or perform calculations on the data it contains." **In a health record, for instance, a heart attack will have the same semantic tag as its more technical description, a myocardial infarction. Now the apple watch shows same.**

Previously, they would have looked like separate medical conditions. Each piece of numerical data, such as the rate of inflation or the number of people killed on the roads, will also get a tag. The data is annexed, tagged and used to the extent of millions of Tetra bytes. It is official we are now perfectly in governments/or hospital hands regarding our health.

U-tube, google-blogging, myspace, and so also, all the web connected technologies which we use to transpire our projects to show case the odds are connected with the digital technology headed by the 5G etc.They are all for 21 st technologies. Also read David Weinberger's book "Everything is Miscellaneous?"

I think you might enjoy it! We have conceptualized that minimum education for everyone, and no one below the age of 16 years shall work. This is the first step for urbanization culture. Hence all family-oriented trades naturally, need to be vocational, and better skill sets needs to be imparted to the individuals. This education is time consuming and is costly for the society as well for the families. That means the society is oriented to pay more costs for same work done by family trained trades man, say about 10 years ago.

As much knowledge sharing is important to all corners of the country, that much it is essential the digital technology has become a guide, a philosopher vital link in the skill development of any given trade. The use of web is consumer oriented. Technologies are students oriented.

Who pays the costs, this is general question. The society redistributes its earnings so that the worker enjoys, not only his life, but also raises a family. How much the worker needs to be paid, is another question, as a technician and as intermediary in the society, he is due 'at least above the minimum wages of a basic worker'.

What skills they need to get, and how much training he needs. This depends on the quality of education the society can impart. If the interactive technologies, based on computers and video films are well made, and if modern equipment are well demonstrated a year of training in each vocational course after his 10 th standard is sufficient. A supervisor needs to undergo another year training on how to manage work force how to compute the work done, and on how to attain safety and quality.

[Piling for a jetty in JamNagar]

Total sequence wise schedule for real time projects needs to be worked out. May be attachment to a industry for 3 months each in all the semesters will enable them to learn to learn construction management, as nothing extra than in degree courses is thought at these Research centers. The name research is for name sake as the staff is not qualified, and do not have any research orientation, forget about producing a viable paper each year.

To tell some real time news, when we were working in one of the companies in 1985 at Alwyn Nissan, [now the company is defunct] Mr. Jaya Bharath Reddy, Managing Director of Alwyn, was coming weekly, to see the physical progress. The physical verification of progress is a prime requirement of the Japanese consultants.

We, and our sub-contractor Choudhary could not mobilize more resources and manpower, as the tender was under quoted. Consequentially internal pressure was building to mobilize more money from completing the project. So the masons were limited to 4 parties working at two locations on a 2 K.M, compound wall. This has led us to show the progress in the morning with four gangs at the entrance of the gate, and mobilize the 'same' masons to factory side in the afternoon, that is where the inspection takes place at the given time.

At one of our project sites our seniors were searching for a urinal. It was start of a project, even the site office is to be made. The pre-fabricated

24

structures were not available [conceptual, as well real time]The boss came out and said, I go to nature as construction man and do not look around for a covered urinal. Our colleges were sitting in the ducts of the wharf deck slab, along with the curing girls after finishing their urinals, and hiding from the works and sun. Now we have readymade offices, and toilets, which can be moved on a Trailor to the required site.

One our colleague was pushed along a kilometer length in his own site, for starting the work without the local leader's approval, in Ponvel, navi Mumbai area, in 1986.

Wit and humor are the most important things in construction. Persistent failures are most common. The goddess, which is the new project, has maximum life of 25 years, of this it runs satisfactorily for maximum 12 years.

The returns are unique. This is more so in case of bridge, and road projects. Same is the case with flats, houses. If the project is delayed, and goes in to cost overrun, then the effective utility time becomes less than 10 years, in these projects. **The Project Management Cycle and its 6 phases...**

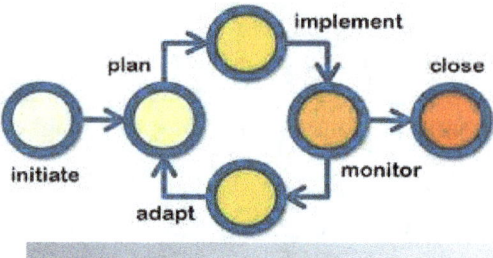

[A place with good natural scenes]

Donors or the owners or even the governments, and societies in common all of them, are requiring better controls and development organizations need more professional managerial competence in the staff responsible

25

for managing projects, it will not be a surprise that in the near future this competence is not only required but demanded.

The management systems have to be created to provide with expert project management consulting and training services based on a methodology customized to the challenges of development projects. Our objective is to serve the needs of our global development community by providing the tools and processes to plan, implement, and monitor projects in a more consistent, reliable and predictable manner, With a methodology based on a project management cycle that incorporates all the processes, tools and practices to effectively manage projects of all sizes.

[The Road users ,in a urban center]

The 21 centuries will mark a big era of new urban centers and new administrative buildings. Many existing, legislature and parliament buildings along with their secretariats have already used the state-of-the-art technology in construction and in utility. Needs for open spaces, new city habitat centers and museums is increasing.

And tsunamis have happened in 2004 December. I was there almost at Hargeisa, a small town in Somali, bordering Ethiopia. On several occasions I had told large audience at these places that JIJIGA, of Ethiopia was a domain, away from main world.

[In Ethiopia, at Somali-Jijigah town in 2004]

It is very difficult to find out who is close to terrorist organizations, or where such are staying. Definitely the head of that organization is a god like person there. I happened to work and see many cities of India, Nepal, Kuwait, Saudi Arabia, Nairobi of Kenya, Addis Ababa, and US cities, Huston, Los angles, Phoenix, etc, in the last 40 years. And this book is a remainder of things to come, as the time passes by, and the population goes on increasing. Climate action, and the heat are two things hunting the humans today.

"Know-how of process Technology" involved.

For an industry it is imperative to know the technology successfully implemented, The Machinery available – and acceptable to pollution norms, Land available at the acceptable place for industry,
Degree of laws governing the area, where the promoter wants to set his industry.
While the above are what an industrialist and team has to conceptualise, the construction teams bill payments depend on the acceptability of the industry by the government.

We have seen big and small industries suffer for want of funds and delayed their projects by few years. The contracting agency working on item rate without a quote for daily machinery, and man power cost, cannot sit idle and pay for all the above.

 Finance: Preliminary report – in terms of similar industries elsewhere showing (a) demand (b) availability of raw materials (c) import of necessary machinery (d) Total cost of plant (d) Time for completion (f) gestation period for Trial runs and market stability. Sometimes even if the product comes out ,but not up to market standards, in some other cases marketing its product by the company is not up to market, in both the cases there will be delay in flow of external finances, and the final bill payment to all contracting companies is abnormally delayed. There has to be a bill finalization and bill discounting mechanism to meet the challenges.

Each one can be explained in volumes, however a proven experience in similar 3 industries with enhance broader outlook, of the trainer.

Consider about the entrepreneur:

a)   Individual, partnership Co., Public borrowing, Private listing or Corporate Management. Follow Company Laws for each of the above, Minimum 15% project cost shall come from promoters, preferably income tax assessors. Construction and setting up: As per Industries act, NBC[National building code], Earth quake resistant, local Panchayat or Municipal laws. Obtaining water, electricity public transport and all other utility services such as bank, Post Office, Police clearance, Lab our license form State and Central agencies, ESI registration where required, Educational institutions, Hospitals proximity and finally fire tenders and installing fire hydraulic systems to meet for insurance cover in longer run of operation.

b)   Checking structural designs, drainage, internal roads, Canteen facilities, workers recreation and proper security. Also installing administrative. Block and accommodating the tax and quality supervisors of Govt. agencies. Weights and measures are also to be taken care of.

CONTRACT CORRESPONDENCE

Most important thing the Project Manger needs to do in any contract is to keep track of the front availability, for each phase of work. The correspondence shall be smooth and shall address in a simple way the record of events, such as handing over site, starting of work, modifications made at site, which will affect either the cost or completion of the project. This includes keeping the records of "Daily Progress Report, Daily Labor Report".

Script the happenings across the region, holidays, strikes etc. Claim the logistical problems where ever they effect the project support. It includes travel of the Workers, Scaffolding or other such support materials. It also quotes about the movement of critical equipment in case of the Total project completion schedule.

Establish **proper laboratory** if it is part of the large contract, and carry out the minimum number of field tests required for each product used in the work.

All the tests done shall be recorded and jointly signed by the contractor's representative and client.

2]Soil tests, Aggregate tests for the road and concrete. Cement and concrete tests as specified in IS 456 and other necessary, contractual codes.

3]Changes modifications shall be confirmed by the client in writing or else the contractor can inform the client the changes ordered at site with in seven days.

4.] Provided the client is aware of financial implications and gives in writing, the contractor will use his right to evolve the extra cost and times involved and inform the client as extra claim in due course of time, as a right.

5] The client shall have no right what so ever to ask the contractor to with draw such a claim unless amicably settled during the tenure of contract. Such extra claims have to be quantified, extra cost evolved (and Administrative sanction to be taken form the competent authority) during the tenancy of the contract by the client. In such event this does not happen then this becomes a claim to be settled either in final bill or during the process of arbitration.

6]However it shall be the attitude of the contractor and his staff to maintain harmonious working conditions with, consultants, clients and of course with sub-contractors. Treating the worker with a human face will yield you at least 10 to 15% extra productivity with absolutely no extra cost. The staff shall always be charged with enthusiasm zeal, strength and a smile on their face (saves accidents).

In the event of a failure of a project, all issues are important for the forensic studies, which are needed for the Judicial review.
Learning few software, algorithms, and use of digital technology is now a need for all.

A High rise Building in construction, with slip form shuttering

CHAPTER FOUR

BUILDING SYSTEMS

Once upon a time there was a painter who had just completed his course under disciple hood of a great painter. This young artist decided to assess his skills. He decided to give his best strokes on the canvass. He took 3 days and painted beautiful scenery. He wanted people's opinion about his caliber and painting skills.

He put his creation at a busy street-crossing. And just down below a board which read-"Gentlemen, I have painted this piece.

"Since I'm new to this profession I might have committed some mistakes in my strokes etc. Please put a cross wherever you see a mistake."

While he came back in the evening to collect his painting he was completely shattered to see that whole canvass was filled with Xs (crosses) and some people had even written their comments on the painting.

Disheartened and broken completely he ran to his master's place and burst into tears. Sobbing and crying he told his master about what happened and showed the pathetic state of his creation which was filled with crosses and correction remarks.

This young artist was breathing heavily and master heard him saying "I'm useless and if this is what I have learnt to paint I'm not worth becoming a painter. People have rejected me completely. I feel like dying"

Master smiled and suggested "My Son, I will prove that you are a great artist and have learnt a flawless painting." Young disciple couldn't believe it and said "I have lost faith in me and I don't think I am good enough. Don't make false hopes. "Do as I say without questioning it. It WILL work." Master interrupted him.

Young artist reluctantly agreed and two days later early morning he presented a replica of his earlier painting to his master. Master took that gracefully and smiled. "Come with me." master said.

They reached the same street-square early morning and displayed the same painting exactly at the same place.

Now master took out another board which read -"Gentlemen, I have painted this piece. Since I'm new to this profession I might have committed some mistakes in my strokes etc. I have put a box with colors and brushes just below. Please do a favor. If you see a mistake, kindly pick up the brush and correct it." Master and disciple walked back home.

They both visited the place same evening. Young painter was surprised to see that actually there was not a single correction done so far. Next day again they visited and found painting remained untouched. They say the painting was kept there for a month for no correction came in!

It is easier to criticize, but difficult to improve. If you want to help people improve their behavior it is worth investing your effort in learning how to help people change their behaviors, attitudes and skills.

Also, always remember not to get carried away or judge yourself by someone else's criticism and feel depressed. Take criticism in your stride; consider that which are genuine and implement those which you think is the best to improve you as a person!!Never compromise. [A Panchatantra story]

The energy efficient, naturally ventilated, green buildings that are of human requirement. Air conditioning and ventilation, or as is called HAVC, shall be the prime target for next age living. Underground constructions may also be sought after by all millionaires. **We are under stress to reduce carbon emissions to have green economy. The fossil fuel industry has to be contained.**

Less requirement for sunlight shall be a demand as the temperatures in cities are rising at an average of 3 degrees per 30 years. So the likely hood of temperatures in Indian cities shall be around 40 to 42 degrees is now prevalent during summer.

Planning a building on ridges, and avoiding valleys is not brought to the notice of the designers and users. Similarly avoiding hill slopes and banks of rivers and ponds is not mentioned. Vaastu adoption for a residential colony has made the well of first owner, to be separated from the neighbor's, septic tank by a few feet. Both were living together, with the contaminated open well water.

The requirements to find a probable slope for drainage of water and a source for supply of drinking water are never thought after in urban housing.

This has left us to find night soil carry pipe laying and treatment techniques, all along the city.

The storm water drainage in developing world is now so costly that the cities find no resources in the years 2020 and above. The instant storms are draining the resources.

Even in the river valley projects the so called ayacut drainage [Irrigation water to be drained out of the lands], or drainage of Surplus irrigation water is to bare necessity and in rains do not augment to the needs.

2. Plan for good ventilation for the building. The forced ventilation is through HVAC. How ever when this fails the scope for natural ventilation shall be kept open. The control over parasites like mosquito has to be thought off while arriving at artificial lighting.

3. Plan for good transport network. Also plan good transport, system to connect all places of daily routine works. There could be when small flying objects may be used for transport and may be helicopters.

Elevated Rail system is implemented in many countries, to avoid, losses due to fire and system failure. Many other future features needs to be thought of in such systems.

TABLE DEVELOPMENT LENGTH OF BARS FOR DIFFERENT STEEL TYPE & DIFFERENT CONCRETE MIXES

$Ld = 0.87$ fy ϕ 4 ζbd $Ld= k$ ϕ

where k is based on character strength of steel, concrete bond stress.

The constant k value for various type of steel and concrete for Tension & compression are given below for reference. Developed length of bar in Tension Type of Value of K for concrete mixes Steel

	M20	M25	M30	M35	M40
Fe 415	47	40	38	33	30
Fe 500	57	49	45	40	36

Fe 550 62 53 50 44 39

Developed length of bar in Compression Type of Value of K for concrete mixes Steel

	M20	M25	M30	M35	M40
Fe 415	38	32	30	26	30
Fe 500	46	39	36	32	36
Fe 550	50	42	40	35	31

The place for takeoff and landing needs special codes to avoid accidents among many such flying vehicles.

The networking of roads, rail transport for fast and efficient transportation of masses needs a careful planning. The services need to run in efficient way. Market utilities, schools for children need to be built in the same area of housing along with the recreation facilities and parks. Good parking facilities for every fourth person at least, are to be planned. 4. The buildings need to be light weight, and no frills, or balconies. Remember a disaster is in pending in Indian peninsula. No area is safe except may be eastern, and western ghats. The disaster can be cyclones, earthquakes, tsunamis, or something not thought off.

While the concept of managing techno commercial aspects has crept in to everyday life, the concept of shifting city centers and building new cities is not new.

CHAPTER FIVE

MAKING OF CEMENT IN CEMENT PLANTS.

The Cement unit comprises of

a) Lime stone Quarry: Line stone is extracted by quarrying, using blasting, and benching techniques and is crushed to 200mm size stones. Conveyor to raw meal storage yard transports this.

b) Morrum or other additives are mined and stacked at the raw meal storage yard.

c) Both the above items are fed into two separate hoppers in the raw meal house where further crushing and mixing is done to the required proportion.

   d) This mixture is stored in raw meal silos. The capacity of silos generally is for a 10-hour production schedule of a day.

Silo Bottom Hopper casting in progress for cement plant

e) This mixture is fed form top, on to a 4 stage or 6 stage pre-heating tower (called pre heater). The pre-heater building is generally 40 Mt. To 100 Mt. Tall and accommodates four stage heaters to heat the mixture to 600 ^{0} C approximately.

f) Thus heated   mixture is fed along with water to a rotary kiln. This kiln is about 2 to 3 Mts. Dia, internally lined with fire bricks and raised at pre-heater end. The slope is towards discharge side. Heating of this mixture with water takes place up to 1650 0 C, where calcinations of lime takes place by addition of water. This is called dry process, and the produce is called clinker.

   g) This clinker is very hot and hence conveyed to stack yard

34

through steel belt.   The clinker is covered against rain. A Gypsum storage yard is located near by. The clinker if grinded and water is added it sets in few minutes. While addition of gypsum gives the time of 30 to 90 minutes initial setting time for cement.

Raw meal silo,and feed hoppers in construction

h)   The clinker and gypsum are fed to two separate hoppers attached to the cement mill where micron size grinding of clinker with addition of about 4% gypsum takes place. This finally grinded cement is stored in cement storage silos. The capacity of silos is equal to the proposed days production.

The conveyor from crusher house to Raw meal silos

35

l) The cement from silos is taken by air slide to packing plant. Here weighing and stitching of cement bags takes place. Conveyor takes these bags to truck loading plat form.

j) Final check of samples before dispatch is done in the lab. While the Excise Dept., records the dispatches from the factory.

CONVEYORS

Conveyors are for lifting loads or materials. In airport the conveyors distribute horizontally and laterally the passenger luggage.

The Design: - Structures are designed for soil bearing capacity, and wind loads. The inclination and length are based on angle of repose of the crushed material to be carried. The end hoppers are designed based on the width of the conveyor and speed of it. That gives total load it delivers to the hoppers per minute.

Also, conveyor elevators and steps are common for mass disposal of passengers. General principals are low speed and versatility in use, safety first.

In industries, the conveyors are used for feeding material to hoppers or silo's for storage and then for mechanization. Such conveyors are made on convex idlers to increase carriage capacity. The inclination of conveyor is based on the angle of repose of the material proposed to be lifted up.

Normal conveyors are nylon threaded rubberized belts resting on idle rollers. At one end a motor rotates the pulley, which in turn makes the belt to pull up wards. The topside of conveyor is called tension side and the bottom side is called the slack side. Adjustable loads are placed on the slack side of belt to keep the belt under tension. This makes the carrying of material easy. Suitable supporting towers for the conveyor, for gallery for inspection and other structures to accommodate motors, drums, hoppers and bins are part of this system.

Apart from belt conveyors, industry also uses bucket elevators (which you can see in villages for lifting water from wells) to move the materials vertically up wards.

There are also what is called air-slides. Pneumatic air is pumped along with the mixture of material vertically as a feed for the industry. Here compressors are used. These compressors have great use in cement concrete grouting, cleaning and sinking of well foundations. Also compressors are used for driving hammer, for piles. The compressors are used to blast hot air in industry where required, and for drilling and blasting purposes.

CORROSION IN CONCRETE:

Corrosion in concrete is common and very effective in rainy areas and near sea. Adequate detailing of rain for cement, use of proper water, proper slump, and complete Vibration/compaction are essential requirements.

Where sulphates are more and tidal variations are high soleplate resistant cement should be used. Also epoxy coating for re-bars is essential for longevity of structures beyond 10 years.

STORAGE OF CEMENT:

Store in air tight, moisture tight, storerooms. Keep at least 0.6 Mts. Above ground with a wooden plat form. Keep air circulation along walls. Do not store nearer to wall to avoid contact of moisture. Not more than 15 bags in height shall be stored.

In case of storage silos of batching plants, the top shall be tightly sealed. In case of stocking for 10/'7 mixture machines, for daily use, use a wooden platform. Stack properly and cover with a thick PVC cloth.

MORTORS: There are used either for masonry construction or for plastering. These can be with lime or cement. Nominal mixes of 1:4, 1:6 etc., are generally in use. Sand should be of good quality free form silt and organic matter.

Proportioning of concrete can be either nominal mix, or Trail mix, or based on (sieve analysis) fineness modulus method. Water cement ratio, and slump shall be based on requirements of use.

15x15x15 cm – 9 cubes shall be cast per each batch of 8 hours concrete and tested at 3 Nos. – 7 days, 3 Nos. – 28 days. This gives a check on the actual site work done and design assumptions.

Pre cast concrete:

Blocks of size.40x20x20 cum. Size are used for main walls. These blocks are made with sand & cement in 1:3 proportions to give 50 Kg/CM2 or higher compressive strength. The wall thickness of these hollow blocks is generally 3 cm.

LIGHT WEIGHT CONCRETE BLOCKS.

These have become the norm now a days because of their size, thin joints, and well developed factory made jointing materials. They are air resistance and keep the temperature to a given level. Construction for interior walls is economical. Out side walls may have gypsum plaster[ready to use factory made] to contain, heat and cold penetration.

Pre-stressed concrete:

Stress in steel is induced before the steel is placed and concreted. A cup and cone arrangement is made where 4 to 6 mm Hard Tensile steel is groups spaced apart. The group of bars is in general 4 to 8 Nos. Then they are kept in the pre cast member where they are to be used. By applying pull force from end (other end fixed) the 'gauge indicates the amount of stress added. Then concrete of M 35 or higher grade (design mix) is concreted. The concrete is cured for 7 days. Usual lifting devices or supports for lifting are used for placing these elements in place.

Lightweight concrete:

Using light and porous aggregate such as coke breeze or slag etc. creates this, (stone aggregate is not used) this has heat-insulating property and also entertains more air, thus becoming porous. This can be used for out side plasters and partitions.

Plum Concrete: This is used as a stabilizing course for water tanks; heavy foundations etc., Boulders of size 200 to 150mm are neatly laid with voids left in thickness of 150mm. Over this 40mm size aggregate 1:3:6 concrete is laid and compacted. This process is repeated to get a thickness of 300mm to 450mm of plum concrete.

Colgraut: This is used specially for monoliths or well foundations. Aggregate of 200mm size is placed up to bottom of sloping portion of kerb. This is about 1 mt. Then grant of cement, sand, water in 1:1 proportion is pumped in side with perforated pipes fixed at 2 to 3 Mt. Distance apart. Finally filling of voids is ascertained through divers and video cameras. The monolith of well is filled for the remaining part with saturated sand and top 300mm is again filled with plum concrete.

water: Since water plays a crucial role in concrete, the water shall have ph value of 8 no sulphates followed. Water in sand and aggregate shall be calculated and deducted from water to be placed while mixing of concrete.

For drinking water IS IS-10500-2012,
water is unfit for drinking if it is bacteriologically contaminated or if chemical contamination exceeds maximum permissible limits1. The maximum permissible limits for various parameters are:Fluoride: >1.5mg/l

Total Dissolved Solids (TDS): >2,000mg/l

Iron: >0.3 mg/l.,Manganese: >0.3 mg/l.,Arsenic: >0.05mg/l

1]Sand- 1540 kgs/m³,
2]Cement-1440kg/m³,
3]Steel- 7850kg/m³,
4]Brick masonry-1600kg/m3,
5]Reinf Concrete-2400Kg/m³
Material testing systems needed for
[1]. Cement-Grade-33/43/53-OPC,PPC,or Sulphate Resistant, or Blast furnace Slag cement.
[a] Intial setting time, final setting time, consistency,[b]Tensile strength[c]compressive strength.
2]. Aggregate sieve analysis, Flakiness Index,Crusing strength
[3]. Sand sieve analysis-gradation
[4]. Concrete mix design
[5]. Slump cone -for workability
[6]. Cubes -compressive strength

[7].Test certificates for cement steel from manufacturer and approved laboratory.

Soil Tests-LL,PL SBC Etc.

Compaction tests for filling. Weights of superimposed loads

Bending moments, Flexure. Deflection for beams and slabs for concrete. For Steel, angles, channels, and beams

Safe stress on steel structures.

Bending moment and shear force, where maximum will occur, formulas., For UDL

Steel structures I_{XX}, Z_{XX} as per Tables in IS.800

CHAPTER SIX

THE SITE TEAM AND BUILDING SERVICES

The Architect or the consultant who prepares working drawings, bill of quantities or schedules, and the specification.

The contractor represented by its Manager (Engineer) with his team, responsible for specifications, drawings and quantities   He is also responsible for quality, safety of personnel (their well being at site) and timely completion of the Project.

The site team established initial grid lines takes the contours (levels) establishes bench marks on site (permanent reference points)

The tem also assess the requirement of manpower, machinery, and peak timework schedule. The tem has to list out and schedule the bought out item and owner supply materials.

The site team also assesses the input requirements terms of contractor's cash flow and payment schedule to contractor. This enables the owner to keep his cash ready for payments on Running Bills, so that the work is not affected.

The Organization has to make the place for workers Colony, batching plant, raw materials storage etc., Also suitable cement go down, stores, medical facilities are needed to be established.

A 'pert' or 'CPM' – same times a bar chart – or for complicated projects – Primavera will held to assess the monthly performance on all fronts.

QUANTITIES:

The Bill of quantities in the most difficult item. The specifications supporting bill of quantities must be simple and detailed, but not exhaustive. If it is exhaustive it leads to litigation and arbitration.

All items of work shall be thoroughly detailed in drawings and in Bill of Quantities.

Water supply lines, bends, elbows, Taps, Rib cocks etc., each shall be priced separately. Similar care shall be taken in drainage and sanitary and in electrical fittings.

IS 17482:2020 on Drinking Water Supply Management System – Requirements for Piped Drinking Water Supply.

IS 10500:2012 on Drinking Water – Specification

Care shall also be taken to detail the class and corresponding IS, BSW or API regulations for fire hydrant and gas line systems. The samples and testing procedure shall be detailed in specifications. For supply materials the payment shall be on FOR destination, rather than part of consignment. All excise materials shall be properly fenced and Dual lock shall be maintained (in accordance with Excise laws).

Testing pressure, working pressure (of 10Kg/cm 2) shall be clearly specified for gas and oil lines. Fire hydrant lines including foam and stand by shall be hazardous free condition for operation.

The stand post hydrant system distances shall suit the insurance agency and the regulating agency.

The IS 15607-2003,along with IS-812,813,817,818 and 822 may be used .This generally covers all structures and pipe lines welding procedures etc.

Due to changes in types and methods of construction, laterally and horizontally, the National Buildings code is revised by the BIS in 2016.It has 12 parts.
SITE TEAM

This team comprises normally owner or the agency building the project, represented by a person who can be an Engineer.

[construction in progrees-1987]

41

Steel structures
RIVETED JOINTS

Always the joint is the weakest part in any structural member. It is necessary up to what strength a joint can be made from 70% to 90%.

In butt jointed – rivets

Normally the dia of rivet shall be $d = 6\sqrt{t}$

Where t is the thickness of plate. Then dia of rivet hole = d + 1.5mm

Strength of a single riveted joint in a butt joint is double shear.

$= 2 \times \frac{\Pi}{4} \times d^2 \times 100/100^3 = KN$ (It is half in single shear)

Efficiency of a riveted joint (n)

= Strength of riveted joint / Strength of solid plate x 100.

Strength of solid plate = area of section of plate x permissible tensile stress.

Yield stress of structural steels. $Fy = 220 – 540$ n/mm 2

Ultimate tensile strength = 1.2 fy.

Factor of safety as per IS 800 = 1.67

Hence, permissible tensile stress comes to 100N/mm 2

The guage is the distance between adjacent rivet lines.

Pitch is the distance between two consecutive rivets, measured along a raw. Edge distance is the distance from edge of plate to center at rivet.

WELDING:

The permissible tress on weld in tension is 150N/mm^2

While in shear is 108N/mm^2

The strength of a weld is equivalent to the thickness of weld, the length of weld multiplied by the tensile stress. That is area of weld x stress. That is area of weld x stress.

That is area of weld x stress.

A factor of safety can always be applied.

Also the thickness of weld in no case can exceed the thinner of the two plates to be welded.

Normal welds specified are – 4mm, 6mm, 8mm and 10mm.

Scaling is to be removed between roots and filling, good grinding is to be done between successive welds to achieve bonding.

The quality electrode material, their burning, the current in transformers either in A.C or D.C. are prominent players.

NDT check, flakiness check, and sonography are done on

a.   Structural materials supplied by the manufacturer.

b.   On welding performed by the site contractor.

For pipelines API Std 1104, **provides requirements for gas and arc welding used in the construction and in-service repair of pipes and components for the compression, pumping and pipeline transmission of crude oil, petroleum products, fuel gases, carbon dioxide and nitrogen.**

The welder qualification, material specification, and the welding procedure, the tests needed, and the qualifications for the personnel to conduct X ray, and other tests are out lined in that.

BUILDING SERVICES:
ELECTRICITY:

Internal concealed Electricity with auto trip 5, 10 & 15 Amps switches for Air conditioner, water heater, Refrigerator with suitable size cables, depending on size of building are required. Cutouts fuse boxes and 16 Amps circuit breakers are necessary. All wiring shall be concealed and one core of wire shall be suitably earthed in a proper earthing pit. Use LED lamps and get the luminous flux approx. equal to 630 per watt.

For construction works it is better to calculate load distribution of welding transformers, motors, pumps, piling rigs, lighting etc., and provide suitable size of 3 ½ core cable where the ½ core is used for earthling.

SEPTIC TANK:

The septic tanks of suitable size with top inlet tank, middle settling tank, and final water outlet tank are to be suitably constructed. This avoids ground water contamination

B2 PILE CAGE LIFTING

[Lifting of Cage for lowering in to pile casing]
Proper lightning arresters and red indicative lamps for night riders are required for high-rise buildings. The size of steel or aluminum flat to be sued and number of lightening arresting pits to be made for building has to be decided by a qualified electrical Engineer (a) based on internal load utility (b) Length of building (c) The light of building.

Transformers of suitable capacity with relay switches shall also be installed to avoid power fluctuations. Load centres shall be established in industrial units for proper load distribution, as well as to enable the monitoring of instruments and functioning of furnaces and conveyors. This monitoring is always connected through load cells to central control Room.

The central electrical control room shall be of proper size of accommodate all cables, cables trays. This room as well as central control room shall have dummy floors for operation purposes. Fire hydrant system of at least two tyres shall be provided. The one from the overhead tank, the other form reservoir, apart from the Co. & other gas cylinders and extinguishers.

DUCTS: Ducts normally are meant to ventilate, circulate fresh air, to heat the rooms, and in tropical climates to maintain room temperature of 280 ^{C}. Ducts are generally made of G.I or aluminum sheets, with rock wool in between the avoid losses in air.

PLUMBING AND WATER SUPPLY:

This job is better handled in India by civil Engineers and plumbers. But in West the mechanical Engineers handle this, thus leaving the floors of baths, rooms, verandahs without a draining out 'P' Trap connection. This causes ample nuisance on the floors.

'S" Traps are used for either Indian or European water closets. 'P' Traps are used at floor drains. All the drain pipes are connection by 3" PVC pipes and led to manhole and then to septic tank.

However it is better that only water closet and urinals are connected separately and with a separate manhole they should be taken to septic tank. All other water shall be led to a soak pit and allowed to recharge the ground water.

The main water lines from overhead tank are 50mm, and from then 40mm and 25mm are drawn for Units of houses. The final bid cock points are of 20mm threaded, concealed G.I. Pipes.

FINISHINGS INSIDE TOILETS:

Shall be with glazed tiles for easy wash and to avoid fungus, bacteria, germs and mosquitoes. There shall be a tower rod. A beautiful mirror, a hand washbasin with a small shelf for keeping all daily use items.

It is preferable to have both a head shower, and a western batch faculty. Suitable soapbox and curtain rods along with W.C. paper holders are to be fixed.

FINISHING FLOORS:

The internal floors are finished with a very slight slope to the point where drainage shall take place. The total slope shall be 12mm and in no case shall exceed 25mm.

Floors are made with slabs made of (a) Granite, (b) marble (c) Glazed tiles (d) or any other tiles or with 20mm chips concrete called Indian patent flooring (IPC). Such IPC floors are useful for laying either rubberized or PVC tiles to give better aesthetic appearance and to reduce noise level. They also do not absorb moisture and keep the floors tidy .

FINISHING KITCHES: This is an important aspect in house finishing. Proper anti termite anti moisture and non-flammable materials shall be used to make kitchen cabinets. All furnishings shall be covered with decolam sheets to make it flame proof.

A modern chimney, an exhaust, fans, a service counter and spacious open area is a must for modern living. An electrical point for a small geyser, a mixture and an oven are also must.

[SLUMP CONE TEST]   .

Proper dish wash and hand wash basins with rubberized auto cleaners for dishes are a must.

Finally the life for cockroaches shall be bad in kitchen, with all dampness and moisture exhausting in minutes.

BED ROOMS

These shall have ample space for a telephone, for a reading table, for a dressing table and not less than a 7ft square Cot to be accommodated.

Suitable cupboards with hangers for cloth and for papers are also to be built by using proper non flammable materials. The finishing of floors can be to suit individual taste.

A

[compound wall with precast slabs is done at site]

[A concrete Mixture of 10/7 in operation]

CHAPTER SEVEN
Teaching in to basics
Rocks and Soils

Kaolin or China Clay: is a hydrated silicate of aluminum $Al_2 O_3 2 S_1 O2 2H_2 O$. It is valuable for Porcelain, china ware, Glazed tiles etc.,

Quartzite: Crystalline rock, white in co lour used widely for road metal because of strength.

Chalk: Pure limestone very soft and easily powdered used for manufacture of cement.

Granular limestone: Co lour is white, soft for crushing, light in weight, and absorbent to water. Wt. Is 2 gm/CM^3, Absorption in 24 hrs? – 4 to 12% when compact can be used for building construction or road metal.

[A inauguration of a construction jetty at Naval Project in 1970 By Director General, Chief Financial Officer and The Chief Engineer.]

Sand Stone: weight 2.25 gr./ CM^3, crushing strength 350 to 400 Kg/CM^2 used for RR, CR and Ashlars work. Durability depends on cementing material, porosity and grain size.

Murum: Is decomposed Literate stone. Good binding material.

Gravel: Rounded water – worn pebbles of any kind of stone. Size around 75mm. Good medium for foundations, filling and Road material.

[A Random Rubble Masonry work in progress]

CERMIC MATERIALS:

Composes of clay products – Like glazed tiles etc., Refractory used as bricks and tiles in industry, Glasses used in building Industry are all ceramic in nature. Clays are essentially hydrous aluminum silicate and are made from decomposition of feldspar rock. Porosity, plasticity, permeability are the measures which account for clay use for making pottery & tiles.

b)   Refractory are a) Acid refractory b) basic refractory c) Natural refractory.

They should stand for high temperatures and do not deform or change in dimensions and should not crack. Their co-efficient of thermal expansion is very high. Acid refractory consists of silica as their base while basic refractories contain either magnetite or dolomite as base.

GLASSES:

Silica is a perfect glass forming material. To lower the melting point of Silica, Sodium Oxide – 25% is added. This has a melting point of 7930 ^{C} and is called sodium disilicate ($Na20$ $2sio2$), however calcium oxide is also added to make the substance stable.

Types of Glasses are:

a)   Soda lime glass with sio_2 – 75%, $Na_2 o$ – 18%

b)   Lead Glass – Sio_2 – 58%, Na_2o – 10%, K_20-10% Pbo – 15 to 40%

c)   Boro silicate – sio_2 – 82%, Na_2o – 3 to 10% B_20^3 – to to 20%

d)   High silica glass – Sio_2 – 93%

LIME: Lime is an oxide of Calcium. Classification based on composition is as follows.
Quick Lime or Fat lime or pure lime.93% by weight of calcium oxide.

Amorphous – (When water is added it burns) and white ($CaO+CO_2$ – $Ca\ CO_3$ as available)
$CaO+H_2O$ – $Ca\ (OH)_2$ on addition of water.
$Ca\ (OH)_2 + CO_2 = Ca\ CO_3 + H_2O$.

Hydrated Lime:
With addition of sufficient qty. of water at manufacturing stage (it slakes – Hydrates – decomposes – splits). This is fine powder and packed and supplied. ($Ca\ (OH)_2$ is composition). This is ready to use for plasters and lime wash and some times (Unimportant works) as mortar.

Hydraulic Lime:
This can be 3 stages of lime
a) Freely or as Hydraulic lime – with 5 to 10% clay.
b) Ordinary Hydraulic lime – 15 to 20% clay.
c) Eminently Hydraulic lime – 20% to 30% clay.
Which is as good as ordinary Portland cement. The Kilns, for burning lime store are either continuous or intermittent.

The manufacture of lime is nothing but calcinatia or burning of materials, in kiln. Heating should be gradual, complete burning is indicated by it bright red co lour. The quality of fuel to be used is to be judged. Over burning or under burning should be avoided. The lime removed can be stocked either in (a) Air (b) as powder (c) Quick lime should be spread as 15 cm layer on a platform and water added to it so that it becomes a paste which can be used for further grinding and use as a mortar.

Properties of Lime:
Slacking or Hydration of lime is defined as the process of chemical combination of lime (Cao) with water so that calcium Hydroxide is obtained

In this process quick lime reacts rapidly with water releasing heat, called heat of hydration and develops an hissing sound, and expands by 2 to 3 times and breaks original lime lumps to fine powder.

Quick lime hydrates in 3 to 4 hours where as hydraulic lime takes one to 3 days. Setting of Lime: Occurs when dehydration takes place or water is lost. $Ca\ (OH)_2$ – $C_2O + H_2O$ [Carbonation) and becomes $CaO + CO_2 = Ca\ CO_3$

The resulting Calcium carbonate is a hard substance acting as a binding material as mortar or as plaster.

Lime is used as: Lime plaster or mortar, Lime concrete, white wash lime sand bricks, Stabilized soil blocks, stabilizing soils for road sub grade.

GYPSUM: Naturally gypsum is available as rock, in dehydrated condition as $Ca\ So_4\ sH_2\ o$ or in anhydrite condition $Ca\ So_4$ and is sedimentary information. Natural Gypsum has impurities such as Silica, alumina, lime and magnesium carbonate. Pure gypsum is white crystalline material and contains – 33.56% Cao 46.51% So_3 & 20.93 H_2o.

Gypsum plasters are binding agents produced by either low or high temperature treatment.

a) Building Gypsum: Produced at $1400 - 170^0$ C by removing Water – in boiling pans and grinding after wards.

b) High strength Gypsum: Plaster of Paris – Produced in autoclaves @ 125^0C. This product is 4 times stronger than the building Gypsum.

The above two are low temperature burned binding agents.

High temperature burnt binding agent – Gypsum is produced after burning to 800 to 1000^0C and then grinding the burned material called OSTRICK – GYPSUM, because it sets slowly and hardness slowly. Accelerators are used, as their setting action is slow.

Gypsum is – Fire resistant, light in weight good insulation, can stick to fibrous Wood. Not used for external plaster. Locks and boards are made. Also used in cement manufacture for retarding initial setting time.

CHAPTER EIGHT
MATERIALS

In Building construction various materials are used depending on (a) availability (b) suitability (c) cost (d) ulterior finishing required.

Bricks: made out of clay are the most preferred in construction. Normal Brick sizes can be around 215mm x 102.5mmx65mm. The flat portion of brick is kept always on the top, when laid in layers. Bricks are laid either (a) English Bond (b) Flemish bond (c) Header bond (d) Stretcher bond. Closers, Bats are used to get break in joints. The joints can be either flush or sunk joints (by 20mm) and thickness not exceeding 20mm.

This is followed it in CR Masonry also.

RR masonry is used in Foundations up to ground, with – keystones x, to see that the walls do not separate, and to a height of not less than 200mm each layer.

Coursed Rubble Masonry is followed – again with keystones to interlock – and give a break in joint – unto plinth level or above that.

Brickwork for ½ Brick walls is in stretcher bond followed by – break up of joints for each course. 6mm m.s. Bars 2 Nos. at 1 mt. Intervals (in height) are provided.

Damp proof course for walls comprising bituminous paint and water proofing cement are laid in layers after crossing final ground level preferably at plinth level. The plinth shall be 600mm above made-up Ground level for permanent buildings and 450mm for semi-permanent buildings. Sand needs to be sieved for mortars on 4.35mm mesh.

Door openings care to be marked and left after plinth construction (Door width + 2 times plaster shall be the clearance). All walls shall be true to plumb. Diagonals shall be perfectly equal. Each layer of brick or stone shall be leveled using thread and spirit level. Tamping rod shall be used to keep vertical and straight.

Lintels above doors and windows shall be of (I) wall width (Breadth) (ii) 2x300 + Length of opening (Length) (iii) Depth not less than the width of the wall or suitably designed.

Lintels can be precast and lifted and placed properly or they can be cast in site, by providing centering and shuttering.

Where chajjas are with lintel they are best cast in place, and supports removed after sufficient weight of wall is built over them, otherwise the chajja will topple and create accidents.

Arches are another form of lintel from esthetic point of view, they are better. A keystone at center is provided in case of masonry arches. End thrust blocks for arches shall be capable of receiving half of the triangular load on each arch. Centering for all arches has to be done carefully and removal shall be from center to corner.

Cavity walls are used in heavy rainfall areas

Hallow blocks of size – 400mm x 200mm x 200mm

400mmx 100mm x 100mm are used.

These are also used for making walls. However care is to be taken to break the using joints of each layer by using half blocks. The walls of these blocks are generally 50mm thick with a compressive strength of 50Kg/CM ^{2} – 1st class and 35Kg/CM ^{2} – for 2nd class equivalent to Brick.

Gypsum and other soil blocks or solid concrete blocks are also in use.

[A architect doing the sculpture work]

Scaffolding:

Generally made of wood or 40 / or 50mm dia MS (2nd class pipes). This works as a platform for masons to do masonry. Double stage scaffolding keeping two posts parallel at 600mm apart is used for plastering & painting purposes. The pipes are tied diagonally and horizontally

Centering:

Centering for RC Slabs and beams can be with wood or steel depending on economy and use. Care has to be taken to see that all supports are perfectly packed at bottom and at top. The supports are to be provided with extra bracing. The load on wood supports for 3M length shall not exceed 500 Kg / CM^2 and for steel pipes 1000 Kg/CM^2. These loads are inclusive of centering material, concrete and live loads.

Centering shall be left in place for 7 days for 3M spans 10 days up to 4.5M spans and 14 days for 6 ^{M} spans.

Normally the finishing works which include a) Plastering b) Flooring c) Painting d) Plumbing e) Timber works eats 50% time and loss of money by way of improper workers doing bad workmanship. Hence, training of the workers and making a model block ready at the 1st instance shall be the practice. (As was done by Shri Ranga Raju Garu of M/s. Nagarjuna Construction Co., Hyderabad.)

Then the estimation of proper cost and time will be easy even with less efficient workers.

Head emitters/coolers are provided below silt of windows. For Euro countries we need to maintain temperatures of
a) Bed Room: 16^0 C to 18^0C, b) Living rooms – 21^0C to 25^0C c) Kitchen: 18^0C Bath – 22^0 C.
Forced (Air) Draughts/Ventilation of 0.25m/Sec. will not be noticed at head level.

CHAPTER NINE

CONSTRUCTION EQUIPMENT

The equipment required depends on the type of project.

1.   **Bull dozer:**   Normally the reverse gear speed is very high. The capacity for work depends on blade width. Clearing jungles, leveling and compacting earth hills it is mostly used, can clear up to 10,000 M ^{2} of height jungle areas for 16 hrs. Work and level about 1000 M ^{2} areas in 10 hours. (From bushes & vegetation and compact the ground that is normally leveled)

2.   **Grader:** With 3.66mm blade will cover 2.74M width effectively generally can cover – 15 cm fill with 4 times compaction about 3000 sq. mts. a day.

3.Road Rollers:

a) 10T – in 8 hours can give an out put of (width of rollers 1.90 M) in formation – 2000 M ^{2}/day in soling – 500 M /day, WB macadam – 250 M ^{2}/day premix carpet – 600 M ^{2} /day.

b) Sheep foot roller drum – width 1.4M at 2.5 KmPh. Can cover at 12 passes on each layer at 200 M ^{2}/hour.

3.   **Tar boiler:** Can give an output of 2700 liters/8hours.

4. **Hydraulic excavators:** Are of 0.9M^3 or Hitachi – 1.1.M^3 bucket capacity can give average – 180 cycles per hour and 150 to   160M^3 / hour production.
    [Figures   are in earlier photos]
5. **Batching Plants:** 100M^3, 50M^3, 30M^3 capacity with cement storage silos are available. Storage space for fine aggregate, coarse aggregate of different sizes and also control room is required. – 10/7 mixers – normally gives 3 M^3/hr. out put of concrete.
6.   **Vibrators:** Plate vibrators, needle vibrators of different sizes (electrical, diesel etc.,) platform vibrators, surface finishing vibrators or trowels and vacuum evaporating   pumps are available. Depending on requirement the types are selected.
7.   **Pavement Breakers**: They are driven with compressed air for 15 Kg. hammers – 35 cfm. compressor is minimum needed. Hacksaw cutting breakers are used for bitumen pavement cutting. And diamond tool cutting machines are used for concrete pavement cutting. Rock drills of various sizes are used for drilling through rock for blasting or separating (with the use of compressed air). The jack hammers are available form 2.5 Kg. – and 2' deep – 1" dia to 30 Kg. – 24' deep and 2 ½" dia Wagon drills are used when dia of bore required is between 3" to 4".
8. **Winches:** Elevators of suitable size are available
9. **Pumps:** of suitable capacity for pumping, storing and curing purposes.
10. **Chisels**   and testing equipment as needed can be obtained. Various chisels for drilling into concrete, cutting embedded rods etc., are available.

11. **Reinforcement Bars:** Clearing and painting apparatus with mechanical bar cutter and bending can be incorporated depending on site requirements.

1) **Spirit level -**   For mason for leveling masonry
2) **Plum bob -**   For masons and for checking verticality.
3) **Thread –** 30M long for aligning masonry and for checking diagonals and checking plastering.
4) **Long Wooden Bar –** For aligning masonry and leveling concrete.
5) **Tamping rod –** 16 dia to 25-dia mm MS bar for making masonry grooves in Hallow block masonry and tamping concrete.
6) **Tape –** 30M and 3 M for measurement.
7) **Water level tube –** 3 M to 10M length for determining floor levels lintel levels, slab levels etc.,
8) **Carpenter tools –** Such as 2 lb and 2 Kg. Hammer, sickle, Nails, split – nail remover, Band saws of different sizes, Nails of different sizes.
9) **Bar bender tools –** Such as 2 Kg., 5Kg. Hammers welded, bending table made for, cutting rails, chisels of various sizes, tying bar and reinforcement.
10) Bar   coating   equipment   where   required.

11) **Survey equipment**
a)   Theodalite
b)   Auto level

These can be used as the work progress for setting the works.

Safety appliances
1) Hand gloves for bar benders, welder
2) Safety shoes for all workers
3) Safety belts for all working above heights
4) Welding glasses and cutting glasses.
5) Protection with hassen wet cloth for all DA cylinders.
6) Current detection and control Breakers including earthling at appropriate places.
7) Dress shall be neat, tidy and no loose frills.
8) Helmets for all workers where work is going on above ground.
9) Protect all over head cables and underground cables from crawlers, cranes etc.,
10) Check the certificate of RTA before using cranes, check their belts, Chains, drums, winch capacity, life of crane etc.,
11) Check all engines for emissions and pollution check all belts, chains for mixer machines and provide guards so that accidents do not happen.
12) Grease & check, drums, check below wheels, keep rear view mirrors, drive below the specified speed for all vehicles (Normally vehicles should be driven below 40 KM. In plant area and 20 KM. At work places)
13) Take precaution to provide protection guard before lifting long pre-cast items and provide for them more than two slings to avoid accidents, also provide guide/control rope at one end.
14) All steel structures on erection shall be bolted properly and secured in position with Guy ropes.

The design of intake wells and supply pipes are checked for Hoop stress (PD/T). The required water hammer devices such as non return valve, Air relief valve, pressure relief valve, sluice valve, Gate valve, foot valve (most of them are diaphragm valves) etc., are to be met in the line for safety. The pipes can be   with internal cement lining or RCC Hume pipes depending on pressure they have to with stand. Suitable, relays, load centers, Transformers, manifolds etc., are to be designed and installed.

The necessary flow diagram for the fluid with the required ROW and other contour maps are to be arrived at.

For Gas and oil lines to maintain pressure, temperature and quantity and to boost pressure where necessary required terminals with 'Scada' and corrosion protection devices for pipe are to be established.

For fire hydrant and foam lines the pump impellers, casing and shaft etc., shall be bronze and PH of fluid shall be checked before finalizing the designs.

CHAPTER TEN

ORGANISATION OF A SITE

The construction site may have to facilitate the following items, either in whole or some parts.

1) **The Project Manager's Office:** Equipped with Telephone, Internet, and local band walkie – Talkies and with charts of staff, cash flow, and machinery.

2) **Conference Room**: Showing the main activities schedule, their monthly/weekly progress. Preferably an audio system and seating arrangements, and a Nova-pan board for conducting seminars etc., Charts showing the safety requirements and mandatory ethics, to incorporate an idea of team togetherness.

3) A Laboratory, comprising of a concrete testing machine, devices for screening of materials. Cube moulds, slump cone, weights and measures. Moisture evaluation apparatus etc., as required. If it is a highway project, compaction testing apparatus, abrasion, and testing of bitumen flash point may also be necessary.

   P.H. value of water can be tested any where, while daily moisture in aggregate, sand, and silt in sand can be found in the lab for adjustment of W/C ratio on day to day basis, cement strength checking apparatus etc., are required.

4) A storeroom for tools and tackles of workmen, including safety devices. A consumable stores, a cement Go- down of suitable storage capacity depending on project (with false flooring made of wooden planks, so that moisture do not creep in are also required).

5) A first aid kit and a supervisor to monitor it and give first aid and give weekly supervision and lectures on 'safety first' are also required.

6) Batching plant area, with a baching plat of suitable capacity, water tank for curing and concreting are also required.

7) Crusher if required is to be located behind and in a non-polluting environmental direction. Preferable to erect

bamboo – or geo membrane curtains so that fines do not flood nearby area.

8)   Suitable workers colony – Isolating skilled, semi skilled and un-skilled separately with suitable water tanks for drinking and usage and toilets at a far away place and general area lighting can also be considered depending on the size of Project.

Please give in a sketch the proposed site facilities for a project you know of, and describe them.

A good CEO clearly spells out his goals and objectives – and then gets his team to come up with ways of achieving them, says Don Mac Rae.

Once the objectives had been described, revised if necessary, and approved by the CEO, we would spend the remainder of the meeting developing criteria for success and implementation plants. [Below Temple Under Construction]

Since this was a time-management seminar, most answers related to the theme of being able to squeeze one more things into a busy life. Taylor listened patiently, then finally said": "No the point is this: If I hadn't put the big rocks in first. I would not have gotten any of the results.

Born in Brazil in 1947, Paulo Coelho is considered the guru of the rich and powerful, and is among the world's most widely read authors. In this rate essay, he comments on leadership, corporate ethics and sustainability. '

For some time now, with ever-greater urgency, more and more people around the world have been asking corporate leaders: "What are you doing? How are you doing it? Why do you do it that way? Have you considered the effects of your actions? What are you doing for the welfare of the communities in which you work?" People feel the impact on their own lives of decisions taken by the leaders of multinational companies and expect answers to their questions.

Many do not seem satisfied with what they see and hear. Could it be that managers cannot answer their questions because they were not expecting them? It is hard to imagine that managers, of all people, are not clear about their objectives and the motives for their actions.

There are many people who believe that corporate leaders have never asked themselves such questions and are interested only in power and profit. However, like most other people, managers too have ethical principles on which to base their actions. The problem is that often they do not know how to apply these principles in practice.

CHAPTER ELEVEN

ROOF; Roofs can be built by using rafters, praline, and sheeting for any unimportant structures.

There are many companies, which offer ready-made steel systems, including trusses.

The most standard practice is to use the RCC slab, designed as per the expected loads. The design and spacing of steel and concrete can be as per IS-456 and other related codes. Use 1200 and its corresponding codes for measurements.

The cement and its properties for OPC,and PPC of,33,43 &53 grades are described in various CODES .Most of the BIS codes are available online for study and down loads.

The estimation of quantities is a simple book, how ever the data books, of the state governments or the CPWD gives the consumption of cement steel, brick etc for various works. The Standard schedule of Rates or the SSR,or the SOR of CPWD gives, the rates for manpower and the material costs at various places[separately].

Use IS;800 in computing steel use[weights]in designs, it gives $I_{xx,and}$ Z_{xx} of the various sections to know the load carry capacity, the bending moment, and the Deflection

FINISHING: Finishing of a roof slab shall be done with in first 3 days of concrete by providing screed not exceeding 50mm at center and at ends 12mm, so that rain water drains happily.

SLIP FORM SHUTTERING.

Hydraulic Jacking system is used to uniformly and slowly lift, a particular shape of shuttering. These shuttering are for unique tall buildings for Silos, Chimneys etc.,

This saves time on labour and also enhances the speed for execution of the tall structures. The loss of timber and the requirement of great manpower over days is avoided, thus giving lots of time saving.

Concreting will be done in uniform layers of 300mm and well vibrated. Layer after layer will be laid for full length of slip form shuttering which normally is 1.5 M height. Of this 300mm will go for overlapping – and positional in line with the bottom layer of concrete. The centreline or center point is connected to all jacks numbering normally 4 to 12 on the periphery. The plumb eccentricity is checked with reference to the central point, which is called spider web. Lifting of slip form is done after an initial setting time of 1-½ hours for 300mm of height at a time. Admixtures are added to reduce the final setting time of 10 hours, to reduce water cement ratio, to get a good slump, and to be able to use concrete pumping for greater heights.

Required insert plates are left for connecting Beams, slabs etc., openings are left where necessary.

STAGING OR SCAFFOLDING:

This is required for concreting of columns, slabs etc., and for masonry plastering, finishing etc.,

Normally staging can be made with single vertical members, called standards spaced at 1.5 M crs. And at 0.6M from the wall. The standards are connected horizontally with ledgers at every 1.5M height. A cross post from ledger to the wall is placed at 0.6M intervals and this is called 'Put log' On top of put logs – planks or sheets are used for men to stand and work. Pipes and clams or the APS units of 1.2 M height are used for the staging. Safety net is of importance around the work place.

For painting purposes double standards are used one at 0.3 M from wall and another at 0.9 M from wall. This is called double staging or double scaffolding.

The development of a method or the detailed planning for construction of a particular project can made-use-of. Economy and cost cutting are important.

Skilled staff of technicians and filed operators are also available reflecting the experience gained with the new equipment of tower cranes, safety nets, and climb forms.

FORM WORK

This is a very important factor in construction. Formwork consists of shuttering plates, made of wood, plywood, or steel. The plates shall have holes or rails for connecting together. In between two shutters either a timber runner or 12mm dia bolts are used.

The formwork after aligning to the required shape of a footing has to be supported form top and at 2/3 ht, in an inclined position. The supports shall be rigid. Remember concrete is in liquid form till it attains its final setting time of 10 hrs. (Initial setting time is 30 minutes to 90 minutes depending on type of cement).

Timber of medium quality of sizes 5 cum. X 5 cm. And sufficiently long (up to 2.4 M) are used for making boards. For tieing plates or boards 7.5cum. x 5 cum. Timber is used. This size is also used for beam bottoms for supporting shuttering. .

The support system for slabs & beams is called centering, while the slab forms are called shutters or shuttering. The casino bellies 9 to 10 cum. are to be braced with bamboos of 2.5 cum. Size at every 1.5 M distance horizontally. The distance of 1.5 M is to be measured from ground to the slab level vertically. Such braced ballies apart can take a maximum load of 1000 Kg. Per each. The ballies placed horizontally at any location shall be within 0.75M..The cantering of beam bottom to be checked for height and uniform level using water tubes and spirit levels. The slab shutters shall be laid to water tightness.

Steel scaffolding and steel props with 60mm dia adjustable jacks are available. These can take much bigger loads of 1.5 T and above and are easy for fixing and removing. The removal of cantering and shuttering is called de-sheltering of the slab.
This shall be done for beams slabs less than 2 M. after – 3 days.
For beams and slabs up to 3 M after – 7 days.

For slabs beyond 4.5M – after 15 days.

For any other hoppers or slabs the designer has to give his specifications. Fallow -IS-456
Mivan or the aluminium form work is mostly used because of its advantages in repetitive uses.

[MIVAN SLAB SHUTTERING]Aluminium

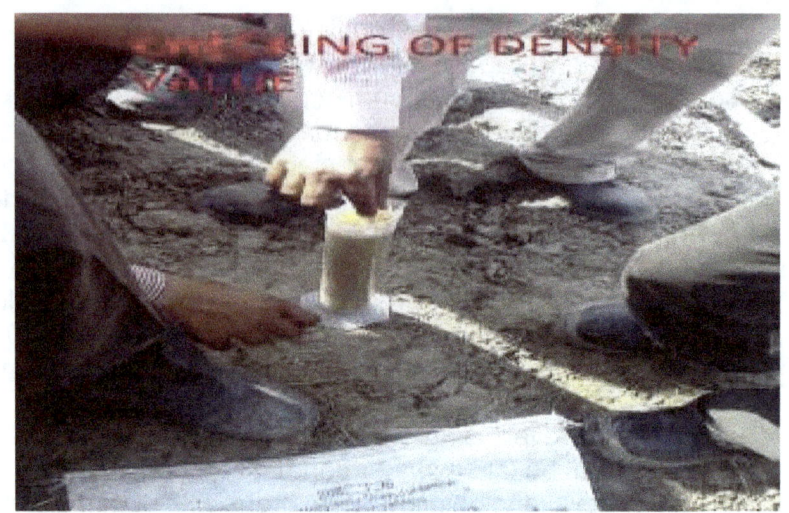
[Finding silt in sand,and bulkage of sand]
It also helps to cast the walls with concrete. The outer walls and
inner   walls   finishing   is   simplified.

[Aluminium side walls form work-MIVAN]

The plaster is thin powder of readymade gypsum or such
materials. With 2 sets in place repetition is almost possible in every
5 or 6 days. The concrete used for walls is High strength using
only 12 mm and down aggregates[specific IS is made for this]

Wide Road with divider in use

TAXES

VAT-Value added tax at 4% is deductable on to total gross value
of the work done including the client's materials supplied freely.
This is need to be paid monthly to the treasury by the contractor
showing the work where the amounts are received.

Where VAT is not followed the following are still under practice.

a) **Octroi:**

It is an entry to exit tax 'form or into a municipal Corporation. '

b) **Excise**

State or Central on manufactured goods, precast concrete items
comes under this, values are to be certified by the owner

c) **Income Tax.**

Generally at 2.22% on turnover is straight –cut.

d) **Turnover tax**

Normally this is called works contract tax at 1 to 2%

(f) Service tax-on services rendered

Customs duty:

On any items used in construction such as plant and
machinery, which are imported, or such items as pipes, Valves,
Steel etc., imported for incorporation in to the works. This tax is
at 40 to 65% depending on type – on the cost of material as per
manufacturer's invoice and to be paid at "port of entry. "

**Municipal Laws: Apart from the taxes it collects, the
municipality has certain regulations on construction of
buildings and industries.**

These include that front wall shall not be more than 1.5M heights.
All round space to be left around the buildings proposed, height of
plinth shall be 600mm above finished ground. No encumbrance

certificate and certificate of registration of land (owner ship certificate) to be provided. Revenue tax on open land, left inside a factory has to be paid annually, to the government.

Water tax has to be paid separately to the municipality or the government.

License fee for construction has to be paid to the industries department, and to the revenue departments separately and then annual taxes are to be paid, to the concerned authorities.

Water lines will be laid after payment of advance money depending on length of pipe and water required. In case of Panchayats director of town & Country Planning has to approve 'lay-outs.'

A Building standing on Jacks above Ground.

CHAPTER TWELVE

Laws in force

Industrial Law:

The overseeing power rests with Inspector of factories, in the District. The guidelines are boilers are to be approved separately. The Administrative. Block and Canteen shall be away and fenced separately from work place. Work place shall be well-ventilated provision of drainage, high ways shall confirm to the relevant National Standards. Power transformers and load centers are to be isolated and fenced. Parking area and entrance for workers shall be provided separately. Front compound shall not be above 1.5 M.

Electricity Laws: The consumption and metering systems are under change. Capitive solar energy plants are encouraged.

Power for construction has to be assessed separately and separate charges will be drawn. The Electrical Inspector (normally one for 3 Districts and SE rank) will inspect and sanction the same. Same in the case with permanent power requirement. Required deposits and consumption charges are to be paid in advance.

It is always better to have a Project report on hand at the starting of a Project, showing staff, machinery and expected profit.

It is always in the interest of consultant to make on the spot 'as built drawings' and submit the same to owner, who in turn can submit the same to the 'inspector of factories'.

Environmental clearance also has to be obtained by the client, during and after completion of works. All chimneys either draught or escape shall be above 100M or above the nearest peak mountain (hill rock) or nearer to sea – River.

The latest CIDC meeting has recommended for changes in construction laws. The basic recommendations are as below. As per the prevailing laws in India, an organization engaged in construction activity requires registration under five different legislations and is subject to inspection by officers appointed under twelve enacted laws.

Further, they are required to obtain licenses under three enactments. All the applicable legislation requires periodical returns and there is a need by construction companies to employ dedicated staff to meet the above-referred requirements under these legislations.

It is also pertinent to note that under these twelve enactments there are twelve authorities who can prosecute or penalize erring construction companies. To deal with the notices issued by different authorities, a team of staff is required to attend the hearings at offices of these authorities.

There is a need to formulate a single law under which one or two Inspectors may undertake the work of inspection under a newly enacted legislation instead of twelve Inspectors under twelve legislations, which is the current position.

The Construction Industry Development Council (CIDC) has mooted the idea of formulating a unified law for the construction industry so that all requirements of different laws may be compiled under one comprehensive legislation and only one Authority be tasked with overseeing compliance of various requirements by construction companies. It is also proposed that the new law prescribe formats for filing returns with the Authority. This Authority should also be empowered to impose penalties for any default in compliance of this law.

Laws of land and basic construction systems

There existed ports, such as kancheeruram, paradeep, machilipatnam bhimilipatnam on east cost of India, when kalinga or pallava kings ruled.

Similarly the history of Alexandria port is as ancient as the shipping industry. However there are no historical details as to how these ports vanished. So the necessity to design and implement infrastructure to with stand long term requirements is of utmost necessity.

Even when we quote the other older civilizations, we are definitely sure that the cities have got dismantled in <u>nature</u> rather than in the <u>wars</u>. We are not discussing in to the safety of a system, or the safety of manpower. *Nature has given so much it can take back some including human beings.*

We need to apply technology, and simple common sense; to find what can be the life of 'steel rails 'that are used for the railways. What calculations are needed to ascertain maximum flood discharage. Can we imagine so much congested track to have a capacity to allow many passenger trains at higher speeds of 80 kms per hour. May be a computer equipped, train needs to run   each   hour   to   find   the   fitness.

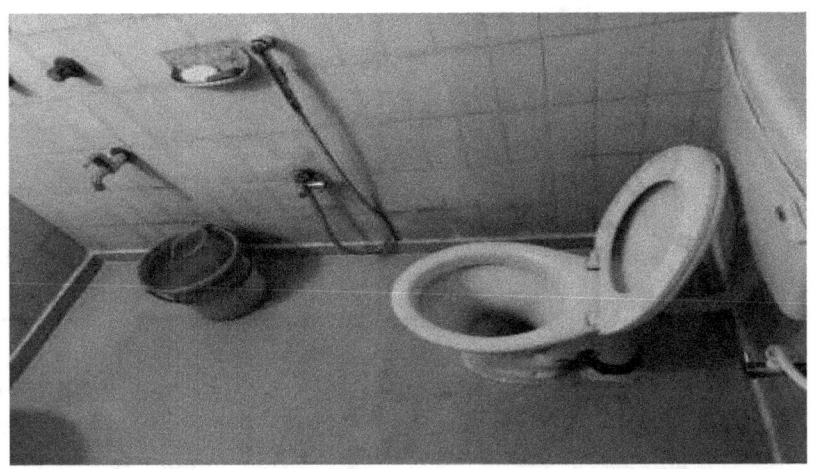

There are no records to show the contracting system followed in the construction of the Cauvery or Godavari dams. Railways in India might be probably the first to engage contractors.

George lake off a great political writer, in New York times best seller' Don't think of an elephant', talks of the political cleverness and wisdom of one party over the other.

The wisdom is to dissect each into practicable, applicable tools. Yes you have planning once you dissect. You know your application, approach, and time frame.

So you are in lead. Another great book of recent times is from Mr. Peter Robinson.

Snapshots from hell [The making of an MBA],is a nice book from the author, who was speech writer to President Mr.Regan's,speech writer. The book is about his experiences at Stanfford.The understanding is that your ability, rather than your pedigree, is important in managing situations.

Drakensburg Mountains are split at many places in Africa, The slope of Nile is so high that any dam, at higher reaches is unstable, uneconomical, and non-functional. However, under great lakes of Nile, in Uganda, Kenya, Zimbabwe, and Ethiopia, there are very vast stretches of irrigable lands. Nairobi, and Addis Ababa, are good traditional financial city capitals, in the east Africa. The other mountains in North Africa are Atlas Mountains are very unstable are responsible for 2004, massive earth quake. These are bordering, Libya, Morocco, and Algeria.

CHAPTER THIRTEEN

It is always in the interest of the site Manger to draw a Project appraisal report, Quoting 'all peaks and achievements' in the completion of the project. This serves the Company to quote for future Jobs based on these statistics. The Indian society for training and development established in 1970,is an affiliate to the international society, runs some courses these are how ever not of practice for construction industry.

Engineering Council of India, New Delhi & Association of Consulting Civil Engineers [India], Bangalore have developed strategies to train the Engineering streams of various disciplines.

Their certificates of competency including internship, Training are value based for various engineering streams in India and Abroad.

Certificate Course

A certificate course for the Trainers [who will be trained either virtually or physically] is intended for the guidance of the researchers, Doctorate students, and the Teachers among all others.

A Certificate needs to be designed for instructions for 4 hours and practical at work or laboratory for 3 hours each day. This system to teach at least 3 or 4 subjects[related] in a month or two will give an advantage for workers

,A six months course including Excell and CAD, will be sufficient for diplamo.Engineering graduates needs 9 months on site training to be an engineer.

With a certification. designed for working professionals, and educators, and corporate trainers, to strengthen their traditional face-to-face education and training skills and improve their ability to work in alternative learning environments, is a need for the society.

These can be flexible and intensive program for individuals seeking professional development or a career change on the whole.

As the need to integrate sophisticated technologies into daily operation grows, professionals or individuals need to understand the interrelationship between instruction and technology. The core idea or the foundation for this certificate can be based on bridging the gap between designing the platform to bridge the theory & design behind construction designs and reports. Project monitoring and reporting etc. The curriculum would consist of three course modules that would be taken as a unit. **Module 1**,2,&3 may be module 4.Some times differed educational system to be encouraged.

The trainers need to be identified and given 3 months training almost every year. The instructions need to focus on
[1] Basic concepts of construction, and real time stories.
[2] laws of land.
[3] Material science-testing, Forensic systems & Quality involved. Basic designs, and the Systems of estimation and costing
[4] Programme and progress verses, outturn and out put verses cost.
[5] Management of laboratory, and supply management verses quality
[6] Management of consumable stores verses billed quantities. Work on theory scrap verses actual scrap for materials.
[7] stores and inventory.
[8] Finances and its management.
[9] Stress management verses new technologies.
Based on his relation to work he may be given addition tutorials for 10 to 15 days.
U tube, Virtual classes have become the norm now a days.
[10] Managing wood work, welding works

[11] Managing pipelines piping and other structural works.
[13] Dam works
[14] managing workers and estimating out turn and training them.
[15] Global scenario, and global law.
1] Hands on experience is a must in operating survey equipment, for surveyors, supervisors, and engineers.
2] Real time projects cycle, in construction is a must. Each project is unique in its nature.
3] Assessing physical, quality of materials and testing them in laboratory at intervals is important.
4] A conceptualized schedule, weekly and monthly needs to be arrived and monitored regularly at each stage of hierarchy.

Born in Brazil in 1947, Paulo Coelho is considered the guru of the rich and powerful, and is among the world's most widely read authors. In this rate essay, he comments on leadership, corporate ethics and sustainability. '

For some time now, with ever-greater urgency, more and more people around the world have been asking corporate leaders: "What are you doing? How are you doing it? Why do you do it that way? Have you considered the effects of your actions? What are you doing for the welfare of the communities in which you work?" People feel the impact on their own lives of decisions taken by the leaders of multinational companies and expect answers to their questions.

Many do not seem satisfied with what they see and hear. Could it be that managers cannot answer their questions because they were not expecting them? It is hard to imagine that managers, of all people, are not clear about their objectives and the motives for their actions.

There are many people who believe that corporate leaders have never asked themselves such questions and are interested only in power and profit. However, like most other people, managers too have ethical principles on which to base their actions. The problem is that often they do not know how to apply these principles in practice.

A wise manager devotes his energy to considering when to listen and when to speak, when to act and when to observe. He should take decisions conscientiously and calmly. As soon as he has taken a decision, he should put doubt aside and unflinching steer the course he has set, even if the conditions he encounters are not the ones he expected.

[An urban river reduced to make a park on its banks]

Reconciling opposites

Top managers face enormous pressures, not only to keep their customers and shareholders happy, but also to attract the best employees and partners, to take the right business decisions and to embrace social responsibility. In many cases they also have to perform tasks that seem diametrically opposed.

They must communicate an at the same time be discreet: they must take and enforce decisions and at the same time take the time and freedom they need to develop new ideas. They must be passionate and it the same time calm.

[Off shore piling in progress]

CHAPTER FORTEEN

WOOD WORKING
TIMBER

Endogenous trees are those which grow end wards like, palm, bamboo, cane etc. Exogenous trees are those which grow outwards like Neem, Mango, Tamarind, etc. Ever green trees are like Deodar, Chir, Fir, Walnut.[B].Broad leaf trees which give hard wood are Sal, Teak, Oak, Beach, Shishum etc. Soft wood, light in weight and color, straight fibers, Distant annual rings, comparatively weak, split easily, good for resisting tensile stress are preferred in construction. Filth Heart wood, sap wood, Bark are to be avoided in usage in any important work.

Seasoning of timber is important [to achieve removal of sap ,to reduce moisture to 8% to increase strength, durability, workability, and it reduces the decay, disease, shrinkage, wrap &crap. Seasoning either artificial or natural ,artificial seasoning is for 3 to 4 weeks, and natural seasoning is almost 2 years.

Tools required for wooden furniture making
Hand tools
i)   Handsaws and backsaws, saws for curve coring, coping saw bow saw.
ii)   Pad saws for cutting starter holes Warrington hammer, wooden mallet, screwdrivers, pincers, chisels, including mortise chisel.
Marking out tools
1)   300mm steel rule
2)   Trisquare
3)   Sliding bevel
4)   Marking gauge
5)   Standard mortise gauge
6)   Hand made large Trisquare
Planes
a)   Jack plane blade
b)   Smoothing plane
c)   Spoke shaves
Boring tools
a)   Bit Braces – Ratchet

b)   Jennings – screw nosed

c)   Flat bit

d)   Counter sink,

Bradawl

a)   Standard

b)   Birdcage.

Clamp heads

a)   Sawing board

b)   Miter box

c)   Mitre block

d)   Shooting board

Sharpening hand tools:

Chipped chisel blade, blunting tools – for all cutting edges are to be maintained sharp.

Grinders of 102mm x 15mm fitted to a drill machine is always used for the purpose.

Horning or applying cutting stone on the oil stone of Carborandum is done, Leather strap, spoke shave – blade holder will save the teeth.

Rubbing it on the oilstone occasionally smooths the scrappers.

Small power tools:

Drill attachments, orbital sander for dual action of smoothening, Jig saws, Router cutters, Band saws (for removing wastes in small cuts)

Ripping tools, push –stick (for against saw) Circular Saws, Jigs for taper sawing, saw bench.

P.S: avoid using wanly wood, or wood with knots or cracks.

Cabinet Hardware & fittings:

Screws of types

a)   Counter sunk head

b)   Round head

c)   Raised head

d)   Twin fast screw caps can be plastic or metallic

Following Hinge is useful

Butt hinge – 2" to 4", Piano hinge to length required Back flap hinge, Table flap hinge, flush hinge, Lay on hinge, concealed hinge, screw – w – Pivot (Ball) hinge, Angle hinge

Catches:

Magnetic, cylindrical magnetic peg lock, Roller catch, ball catch, double roller catch, Ball catch.

Shelf fittings:

Ring support, self-stud, cylindrical fittings
Locks:
Cupboard lock, till lock flap lock
Knock Down fittings:
Back joint fitting, dowel screw – connecting screw, Tee nut & screw socket, and Hanger Bolt and corner place.
Plates and Brackets:
Expansion plates to avoid shrinkage and expansion and crack, corner bracket and angel, wardrobe materials like rails and supports.
Jointing:
i)   Lab joint
        a) Full
        b) Slope
        c) Side slope
        d) Angle to sides
ii)   Mortise and Tenon Joints
iii)   Bridle joint
iv)   Housing /half lap joints in furniture main boars (to accommodate the cross bards)
v)   Dovetail Joint
vi)   Butted joints.

CHAPTER FIFTEEN

CRIME DETECTION and FORENSIC ENGINEERING

For that, what is need is to identify the criminal investigation system, and ways and means of detecting the motive and what makes one to go that far and the systems which are needed for punishing the greedy needs be placed in the legal system.

Greediness of the entrepreneur or the others in the society leads to an imbalance in the working environment and the social culture of the nation. All these things happened in 17 th and 18 th century in the developed world, when there was no policing, and when there were no jails. How ever a nation cannot label more citizens as criminals and keep them in jails. This again is a loss to the society; hence a balanced approach is needed in handling the offences against the citizens and the Nation.

'The evolution of new civilization started with the 1]. Trail based Justice system. Then the trials were only witness and testified. Authors wrote books, of 'Perry Mason" etc,and 'Arthor cannon Doyle' wrote about crime investigation.

How ever it all started when empty waste bottles were sent with the insurance from US to Europe.

Now the engineer has the functions of 'detecting the materials', used their 'testing process' and 'procedure used' and the compositions made in attaining the structural safety and construction tequinques.

The engineers, needs, material certificates, including suppliers' names in his register.

The 'tests carried' and how they reflect on the life of the structure. Having gone through the test corticates, of cement, steel, by the producer, by his own lab and third party, he has to 'chronicalise the construction sequence'.

Then design his concrete mixes, the pour dates of concrete, and the slump, and the cubes crushing. That system is needed for all structures of concrete.

Other finishing and masonry, aluminum and finishing on roof tops needs to be obtained for the manufacturers test and guarantee.

Mix designs are most important, **like steel, testing in steel structures. The steel structures need the tensile strength,**

weld strength, welder qualification.
Apart the assembly has to with stand during construction also, its own load carrying with dynamic and wind loads is important.' **We remember a silo construction during a cyclone period, which has large effect on the down pour. The wind loads during the same time have made life uneasy for the construction managers.**
The "collapse" of civilizations, it was clear that this is another example where "Collapse" has specific implication of "imploding" under its own weight or mismanagement or something. While the Classic Maya may have "collapsed," the Post-Classic Maya were conquered by the Spanish, and had their monuments destroyed or forced into neglect. Even then, to have a civilization conquered is not necessarily to have it ended.
In ancient times many large monuments structures had been built across the globe. Thought to be build by different cultures, there are certain similarities between them which could prove there ever was a global civilization or organization. in the ancient past.
India too saw many structures built, some destroyed, some partly available. But where nature has played its role is not much recognizable. The forensic systems are the only way to find when built and when vanished.
There are many mysteries considering the construction of old structures. Many were build using very large and very heavy stones. How could they have lifted them in ancient times without the aid of machines? Why did they go into the trouble of dealing with such heavy stones as they could use lighter and smaller stones.

By 1500 BCE, the once vast and powerful civilization began to decline at some point it suddenly ended.This could include some quakes in Indian seas also Historians are uncertain why this area's power declined. There are some theories that a great earthquake crumbled cities and changed the path of rivers, which caused them to move to a new location. Another theory claims the climate may have changed, which forced them to move. Yet another theory suggests invading armies destroyed some cities and forced most people to move. One thing we know for sure is that the civilization that once lived in this area ended and new people moved into this area.

There have been great discussions on the parallel between the destruction of old civilizations and the present systems by modern authors and Anthropologists in the book "The Historical Evolution of world systems "Edited by Cristopher Chase and E.N. Anderson.

Please think why machines are needed for the Civil engineering works.
It is like why doctors need a certificate from Xray.
Yes X ray was developed for material inspections also.
Also we have the teeths and jaw made us develop a jaw crusher.
A folded hand is symbolic of our excavator.
The cocktail mixtures, mechanical and pharmaceutical are nothing but our stomach, which has seven layers of skin, to inject the needed acids, produced elsewhere.
Two things first-we had the machines, we have signal system, we have sensors.
We developed functional machines, based on human evolution and working.
An extension is the Artificial intelligence.
Learning a computer, a language, digital infrastructure, and algorithm is easy and a necessity.

[USA House construction]

CHAPTER SIXTEEN

Mix designs and testing procedures:

Concrete: Air-ENTRAINING ADMIXTURES – These admixtures cause air to be incorporated in the form of minute bubbles in the concrete during mixing, usually to increase workability and resistance to freezing and thawing. They control the amount of air in fresh concrete and disperse properly sized air bubbles throughout the concrete. The origins of Air-entraining admixtures are as follows.

a.   Natural wood resins:

b.   Animal or vegetable fats and oils

c.   Various wetting agents, such as alkali salts of sulphated and sulphonated organic compounds,

d.   Water soluble soaps of resin acids and animal or vegetable fatty acids, and

e.   Miscellaneous material, such as sodium salts of petroleum sulfonic acids, hydrogen peroxide, aluminium powder etc.,

The entrained air bubbles (approximately 0.05 to 1.25mm dia) reduce the capillary forces (the force causing absorption of water by concrete) by restricting the effective length of each capillary pore in concrete. The capillaries are interrupted by relatively large air voids in air-entrained concrete. The voids cannot fill with water from the capillaries because of surface tension effects and, therefore, under freezing conditions, they behave as 'expansion chambers' to accommodate the ice formed.

Entrainment of small amount of air results in concrete of insufficient durability, whereas with large amount of air entrainment there is an excessive strength reduction in concrete. Therefore, an optimum percentage of air giving a balance between compressive strength and durability must be used in practice. Table 190 gives optimum air contents or concretes of different maximum sizes of aggregate.

INFORMATION ON ADMIXTURES – To facilitate approval of an admixture, the following information is needed:

a.   The trade name of admixture, its source, and the manufacturer's recommended method of sue:

b.   Typical dosage rates and possible detrimental effects of under and over dosage:

c.   Whether compounds likely to cause corrosion of reinforcement or deterioration of concrete (such as those containing chloride in any form as an active ingredient) are present and if so, the chloride ions by mass of expressed as equivalent anhydrous calcium chloride by mass of admixtures; and

d.   The average expected air content of freshly mixed concrete containing an admixture, which causes air to be entrained when used at the manufacturer's recommended rate of dosage.

CONCRETE MIX DESIGN: Entrapped air in concrete is assumed as follows .;

Aggregate size in mm	Air on % volume of concrete
10.	3.0
20.	2.0
40	1.0

Approximate sand and water required.

For W/c = 0.6 & workability = 0.8 CF up to M-35 grade.

Size	Water/M³ of concrete In liters	Sand as percent of Total aggregate by Volume
10mm	208	40
20mm	186	35
40mm	165	30

Standard deviation for mixes with different degree of control

Mix	Very good	good	Fair
M10	2.0	2.3	3.3
M15	2.5	3.5	4.5
M20	3.6	4.6	5.6
M25	4.3	5.3	6.3

Design procedure:

28 day characteristic strength required   = $20 N/mm^2$

Size of aggregate – 20mm (mixed say 20 - 60% 10mm – 40%

Degree of workability = 0.90 (compacting factor)

Quality control – Good

Type of exposure   - Mild

Specific gravity of cement – 3.15

Specific gravity of coarse aggregate – 2.60

Specific gravity of fine aggregate   - 2.60

Water absorption – CA – 0.5 percent – FA –

Free moisture in CA – Nil, FA – 20 percent.

F.A confirming to Zone-III of IS: 383 Table 4

Target mean strength of concrete for at=1.65 20+4.6x1.65 = 27.6 N/mm^2

General W/c – as per tables = 0.50

(Less than 0.65 prescribed from tables for mild expose)

As per above tables water required = 186 Kg. Sand 35%

For change of W/C from 0.6 to 0.5 and sand of Zone-III

Increase water +30% and sand – 3.5%

Sand as 1% volume is 31.5%

Water = 186 + 5.58 = 191.6 liters

Cement = 191.6/0.5 = 383 Kg/m^3

Maximum cement required for mild exposure for RCC = 250kg/m^3

Maximum cement required for moderate exposure for RCC =290kg/m^3

Maximum cement required for severe exposure for RCC =360Kg/m^3

And maximum W/c ratio is 0.65, 0.55, and 0.45 respectively.

Hence 1-0.02 = 0.98M^3 = (191.6+383/3.15+1/0.315. fa/2.60) X 1/1000 =

546 kg/M^3

And 0.98M^3 = (191.6+383/3.15+1/0.685. ca/2.60) X 1/1000 =1187 kg/M^3

The mix proportion then becomes

Water	Cement	F.A	C.A & 2
191.6	383	546	1187
0.5:	1:	1.42:	3.09
0.5:	1.42:	1.42:	(1.854: 1.236)

Calculate for 1 bag – 50 Kgs – requirement also calculate yield.

Calculate adjustment in water for FA & CA.

Comparison of rocks for use in concrete and as road metal (in Abrasion).

Rock type	Crushing strength MN/M^2	Crushing Value	Abrasion Value	Impact Value	Attrition Value		Water Absorption
					Dry	Wet	

Basalt	200	12	17.6	16	3.3	5.5	0.1 to 0.3
Granite	185	20	18.7	13	2.9	3.2	0.2 – 0.5
Lime stone	165	24	16.5	9	4.3	7.8	0.2 – 0.6
Quartzite	330	16	18.9.	16	2.5	3.0	0.2 – 0.5

Basalt & Quartzite are well suited for road metal and Granite and limestone for concrete.

CHAPTER SEVENTEEN

WATER

Introduction : The requirement for this item in human civilization is indespensible.The governments have to rationalize the supply and collect user charges for this. The private agencies do filter and bottling for human drinking and cooking. The earth and water absorb all human excreta, and even human Skelton. Still the earth yields food grain, and the rivers carry water for living. It is good to have the knowledge about water and its purifications for human drinking and industries.

The need is to plan about preserving rain water, storage, and supply. Hence the need to rationalize the city water supply .In the beginning of creation there was water only on the earth Water is said to be our life. It is indispensable for plants and animals also. It is an essential to maintain life line of agriculture and industries. Water is as important as air.

Source of Water:

Water is tremendously obtained on the earth surface in sea, river, lake, and pond. Underneath the earth surface it is in the well.Spring water comes out from the earth.

Earth's crust contains porous and non-porous rocks. The porous stratum percolates Water and non-porous Strata holds the water underneath the ground. Water is found in the form of ice on the mountains, in the form of clouds in the sky. Rain fall replenishes water. It is obtained by the chemical combination of hydrogen and oxygen.

Normally rain gangue stations are established to measure the amount of rainfall in a given day. The gangue, the area are all standardized. These rain Ganges are like a network. Each rain gangue covers some area in square kilometers. Over the year the total daily/monthly rain fall in the area is established. This gives over rainy or flooding years. This also gives average rainfall per year in the area, and the drought years.

Califluculater

Normally, where crops are gown with rainwater, they require 50 cm. To 100 cm. Rainfall in 60 to 90 days. This is true for most of dry crops, except rice, wheat, sugar cane etc., Rice requires about 2 M water in 120 days and sugar cane about 3 M water in 240 days. Since in most of the area the monsoon, rainfall is on an average 1 M, the remaining is supplemented by storage reservoirs, Dams etc.,
The rain fall in given area partly goes in to ground water as per collation part of the rainwater gets evaporated.   Only about 80% of rainfall is available as flowing water – called run off.
 Since, this flow is seasonal, it is not possible to store this water entirely, and hence rivers carry water to Oceans.

Impurities in Water

Perfectly pure water cannot be obtained even by the most elaborate technique. It always contains impurities. The purest known natural water is rain water; however it contains of organic matter and dissolved gases, mainly oxygen and carbon dioxide taken from air.

[Uv light-water treatment]

The impurities commonly found in natural water may be broadly classified as follows:
Suspended impurities. Mud, sand, vegetable matter, sewage, trade waste and bacteria.Dissolved impurities. Compounds (bicarbonates, soleplates, chlorides and nitrates of calcium , sodium , and magnesium , iron and manganese compounds, silica alumina , sewage and various trade wastes, and gases-oxygen , carbon .Carbon dioxide , Hydrogen sulphide and nitrogen. Rain becomes saturated as it falls.Rain also contains small amounts of nitric acid formed during electric storms and of sulfuric acid formed from the sulphur of coal during the combustion process.

The water of lakes and rivers consists partly of the rain that has fallen directly on the surface of the water, partly of water that has been in contact only with the surface of the soil, and partly of water that has percolated through the soil and has opportunity to dissolve the soil constituents.

Rain water that percolates through cultivated ground dissolves organic matter usually oxidized by bacteria, when the water reaches a depth of one meter, the organic matter is completely oxidized and the percentage of CO_2, increases. As the water percolates further through the soil, it continues to dissolve mineral matter. The underground water flows through gravel and other porous strata and may be ultimately come to the surface from a well. The amount of dissolved solids will usually be larger than is the case with surface water. Organic matter, such as that contained in sewage leaking from a cesspool or sewer several metres below the surface may travel underground for long distances without complete oxidation and without elimination of the bacteria it contains. Therefore the water from wells and springs is not considered safe for drinking unless it is examined by laboratory tests.

Nitrogen may be present in water in the following forms: (I) free ammonia, (ii) albuminoid ammonia, (iii) nitrides and (iv) nitrades. Free or albuminoid ammonia is usually considered as an evidence of insanitory pollution.

B, coli or colon bacilli or coliforms are harmless becteria present in water Their presence in a sample of water indicates pollution.Chlorides of sodium and calcium impart salinity. Sulphides of calcium and magnesium causes hardness of water.

CHAPTER EIGHTEEN

The necessary flow diagram for the fluid with the required ROW and other contour maps are to be arrived at.

For Gas and oil lines to maintain pressure, temperature and quantity and to boost pressure where necessary required terminals with 'Scada' and corrosion protection devices for pipe are to be established.
For fire hydrant and foam lines the pump impellers, casing and shaft etc., shall be bronze and PH of fluid shall be checked before finalizing the designs.

Water Analysis
There is plenty of water but perfectly pure water is not available anywhere. Water consists inorganic and organic substances in suspension, as well as bacterial and biological life manifested by fell diseases like cholera, typhoid, etc. In order to determine all these impurities, the water is analyzed and treated accordingly to make it suitable for domestic, industrial and agricultural uses. The analysis includes determination of the following :

1.Total solid., Suspended and dissolved organic and inorganic solids, like mud, sand, bacterial, coli form. It is expressed in parts per million (ppm) parts of water.

2 Oxigen.Dissolved oxygen is an indication of amount of organic matter, mainly derived from the decay of animal and vegetable sources.

3.Nitrogen.Nitrates of calcium, sodium, magnesium impart pollution, when the amount exceeds 5 ppm.

4.Amonia.Dissolved ammonia is an indication of pollution by

90

sewage.

5.Chloride. Chlorides of calcium, sodium impart salinity.

6.Sulphur.Sulphates of calcium and magnesium causes hardness of water.

7. Hardness.Hardness of water is due to the presence of dissolved calcium and magnesium bicarbonates, chlorides, nitrates and sulphates of calcium and magnesium. It is expressed in parts per million (ppm) by weight in terms of calcium carbonate.

1 ppm = 1 mg/litre of gm/m3

Soft or hard water is specified on the basis of ppm as :

Soft water	=	0 - 0.55 ppm.
Hard water	=	0.56- 200 ppm
Very hard water	=	201- 500 ppm

Soft water readily forms lather with soap. Hard water does not readily forms lather with soap. Ordinarily, on the soap consuming capacity the standard of measurement is degree clark, hardness, degree referring to grains per gallon. The hardness of water according to this standard is :

Soft water	<	10°
Medium hard water		10°--20°
Hard water		20°-- 30°
Very hard water	>	30°

One degree of clark hardness may be equated to 17.1 ppm.

8 p H Value.   PH is defined as negative logarithm of the hydrogen ion activity of a medium, expressed as :

$$pH = - \log (H+).$$

It is a measure of alkalinity or acidity of a solution. Water with pH = 7 indicates neutral solution. Water with a pH of more than 7 is alkaline and is less likely to attack matals. Water with a pH of less than 7 is acidic and may corrode metals unless acidity is neutralised by treatment.

Reports of Water Analysis

The actual amount of dissolved solids in water is small as compared with the water itself, however, it may cause considerable trouble in a boiler. The total solid in bad water may only 0.03 or 0.04 percent, which shows approximately 99.97 percent of water, that might lead one to believe that the water is essentially pure. In order to avoid showing percentages in the third or fourth place of decimals, water analysis are generally reported in parts per million by weight (ppm). Some analysis are reported in grains per gallon, especially in British litrature. One grain per U.S. gallon is 17'1 ppm, and one grain per British gallon is 14'3 ppm. The use of parts per million (ppm) is preferred because it is easy in making calculations.

If many calculations of water treatment are to be made, it will be found convenient to report water analyses in equivalents per million parts by weight (epm), since one equivalent of one substance will react with or replace one equivalent of any other substance. To convert ppm to epm, divide the ppm of the given substance by its equivalent weight.

Fresh water having high sodium and chloride has been contaminated with sea water. Rain water is considered to be the purest form of water. Hard water contains dissolved calcium, magnesium, iron and aluminum compounds. This hardness has probably been acquired by percolation through beds of limestone and gypsum. Nitrate content in moderately hard water is usually high. In low hardness water of streams, the carbonate hardness is usually lower than the non-carbonate hardness.

Water containing less than 150 ppm of hardness are generally classed as good, those containing from 150 to 350 ppm as fair, and those exceeding 350 ppm as bad.It is to be noted that the calcium content of fresh water is high relative to the magnesium content, it is reverse in sea water. Fresh water seldom contain very much alkali metals or chlorides.

TABLE 14.1   Analysis of Water of a Lake in Parts per Million

Turbidity	2	
Silica (SiO_2)	7.4	
Iron (Fe)		0.06
Calcium (Ca)	13	
Magnesium (mg)		3.1
Sodium and Potasium (Na + k)		32
Carbonate (CO_3)		0.6
Bicarbonate (HCO_3)	5.6	
Sulphate (SO_4)		2.1
Nitrate (NO_3)	0.5	
Chloride (Cl)	1.1	
Total dissolved solids	60	

Water Treatment

Water treatment consists of removing some or all impurities to make the water suitable for domestic and other specific purposes.

Water treatment includes filtration, sterilization and softening. Municipal water conditioning is very often done for producing water suitable for drinking and domestic purposes. The object of purification for domestic supply is to remove turbidity, co lour, taste, odors and also bacteria which may cause disease. The basic system of water treatment may be as follows :

1. Sedimentation of suspended matter having specific gravity more than one, by gravity. It consists of allowing the water to remain at rest in setting tank so that most of the suspended impurities settle to the bottom. If the suspended particles are very slow to settle, coagulation is done which consists of adding to the water in its passage to the settling tank, some coagulant (usually alum* or aluminum sulphate in solution). This forms flocks, i.e light, wooly-looking insoluble precipitates, to which colour-producing colloids, bacteria and very fine particles adhere and quickly settle down, and are thus removed from the water.

2. Filtration is done either on slow sand filters or on rapid sand filters.

Slow sand filter consists of a bed of sand about 1 m thick through which water is passed at low rates. 5000 to 10000 klitres/m2 of filter bed per day. The size of sand particles in the filter bed is 0.4 to 0.5 mm. Filtered water is collected through perforated pipe.

Rapid sand filter or mechanized filter consists of 60 cm - 75 cm thick top layer of sand (size of sand being 0.4 - 0.55 mm), 25 cm - 45 cm bottom layer of gravel (size varying 40 mm at bottom to 3 mm at top) resting above an under drain system to collect the filtered water. Rapid sand filters can operate 500 000 liters / m2 per day. The high rate of filtration is made possible by (I) coagulation and sedimentation ahead of filtration, (ii) using family coarse sand, and (iii) backwashing the sand every alternate day to keep (he bed clean. Since the area of beds required for such filter is much less than that required for slow sand filters, rapid sand filters are now installed every where for large scale purification of water.

3.De-aeration is done either by de-aerator or open heater to removed dissolved oxygen, nitrogen, carbon dioxide in air.

4. Aeration of water is done by exposing the water to the atmosphere for the following two purposes :

(i) Mechanical reduction or removal of gaseous or volatile impurities such as CO_2, H_2S.

(cml) Supplying oxygen necessary for chemical

oxidation of impurities such as ferrous iron, manganese and removal by setting and filtration.

5. De-mineralization or de-ionization is done to remove mineral matter by cat ion exchange followed by anion exchange.

6. Distillation is done by vaporization and condensation for making the water free from any dissolved solids.

7. Flocculation is done by an integral unit clari-flocculator to remove flock. Raw water enters a central chamber in which chemicals such as aluminum sulphate, ferric chloride, ferric sulphate are fed and circulated by agitation. Precipitated flock and water with the turbidity pass to clarification chamber., The effluent pass upwards through a blanket or suspended flock.

8. Sterilisat5ion of water is done for destroying disease-carrying germs by chlorination of ozone. Chlorination is the treatment with chlorine and ammonia for sterilisation, prevention of growth of algal matter, removal of taste and odour defects.

.Softening. Hard water is made soft, by external treatment in a separate plant or by internal treatment by addition of chemicals.

10. Water for boiler and steam condensing system is conditioned by treatment with amine, to protect the boiler feed system and to meet the operational requirements.

Hard and Soft Water :

Water that contains little dissolved solids (compounds of ;calcium, mangnesium, iron and aluminium) has traditionally been considered soft, and that which contains much dissolved solids has been called hard. The origin of these terms is probably due to the behavior of water with soap. Water that readily forms a lather with soap is called soft water. Water that does not readily forms a lather with soap is called hard water. Hardness of water has been defined as the capacity of water for added to precipitate all the metal ions producing hardness. This is the basis of all the well-known soap method for determining hardness. The total hardness $CaCO_3$ in ppm is fifty times the sum of the equivalents per million of Ca, Mg, Fe and A1, since the equivalent weight of $CaCO_3$ is 50.

Spring water is often hard, sea water is always hard, deep-well water is sometimes found to be hard.

. Temporary and Permanent Hardness

The amount of hardness equivalent to the sum of the carbonate and bicarbonate ions is called carbonate hardness, and the amount of hardness in excess of this is called non carbonate hardness. This method of subdivision is preferable to the older way of dividing the hardness into temporary and permanent hardness since the latter method cannot be definitely interpreted in terms of the chemical species existing in the water.

Temporary hardness is caused by the presence of dissolved bicarbonates of calcium and magnesium. It can be removed by either boiling the water or treating it with hydrated lime.

Permanent hardness is caused by the presence of dissolved chlorides, nitrates and sulphates of calcium and magnesium. It cannot be removed by mere boiling or adding lime. precipitating soap. The dissolved solids react with ordinary soluble soaps to form insoluble, sticky soaps of the dissolved solids. The soluble soap will not form a lather until a sufficient quantity has been

CHAPTER NINETEEN

THE PRINCIPLES & OBJECTS OF CHLORINATION OF WATER

01.   Oxidize all organic matter.
02.   Sterilize, that is, destroy all disease bearing germs.
03.   Leave a little surplus or residual chlorine to act as a safeguard against further contamination but not so much as to cause taste or other troubles.

Although cylinders contain 60 to 70 lbs. Of liquid chlorine, not more than half the quantity be drawn from the cylinder in 24 hours, as the rate of evaporation of the liquid within the cylinder causes a lowering of temperature and if the gas is drawn off too rapidly a covering of frost, will appear round the cylinder on the outside of the container, resulting ultimately in the contents freezing solid.

Therefore where chlorine of more than 20 lbs. Per day is required, a cylinder of 70 lbs. Should be used for each 20 lbs. Since the rate of more than 1 lb/hr. from one cylinder at Indian temp 80° F & above will cause problem.

In cold climate chlorination room temp should never be allowed to go below 40° F. Ventilation in chlorine room should be at the bottom of G.L.One lb. Of liquid chlorine at about 68° F will produce 5cft. Of gas and if evaporated 50,000 cft. Of moist air will cause severe caughing.

Break Point Chlorination :

On application of Cl2, the residual chlorine content reaches to a certain peak value & then with more application of cl2, the reduction in residual cl2, is noticed and this is continued until a minimum residual Cl2 value is reached, after which only the Cl2 content value increase in proportion to the applied dose. This minimum residual is known as the BREAK POINT chlorination & the plotted curve is known as Break Point curve.

Minimum contact period for Cl2 - 15 to 30 minutes and after 30 minutes residual Cl2 should be atleast 0.2 P.P.M. Recommended Dosages are as under. :

Recommended dosages of Cl2

1. Filtered water supply   3 to 3.5 lbs./per million   gallons
   i.e. 0.3 to 0.35 P.P.M.
2. Uncontaminated surface   5lbs./per million gallons i.e. water

for Domestic   0.5 P.P.M.purposes.

PH Value :

The PH value at which waters do not become corrosive is known as the cquilibrium or saturation PH. The difference between the actual PH value and the equilibrium PH is known as the "LANGELIER INDEX". Water which has a positive Langelier Index is likely to deposit $CaCO_3$ but if it has a negative Index then it will dissolve $CaCO_3$ and will be highly corrosive.

Alum & Lime Dosings :

10 lbs. Of Aluminium Sulphate $Al_2 (SO_4)_3$ $18H_2O$ contains 5.13 lbs. Of anhydrous Aluminium Sulphate & 4.87 lbs water and the equivalent of 1.53 lbs of alumina or aluminium oxide which theoretically can react to produce 2.34 lbs. Of insoluble aluminium hydroxide (the main constituent of so called ' Alum Floc').

1 lb of CaO to 1000 gallons of water assuming it to be at 50° F (Saturation figure for the lime as CaO) is only 92.12 grain/gals.

Solubility of like in water decreases with increasing temperature from 1000 grains/gallon at 32°F to 39.6 grains/gallon at 212°F.

1 part of hydrated lime will neutralize 1.19 parts of freee CO_2, producing calcium bicarbonate $Ca(HCO_3)_2$ equivalent 1.35 parts of hardness as $CaCO_3$.

Twice this quantity of like will be required to neutralize the free CO_2 and to precipitate it as chalk or $CaCO_3$ otherwise the lime increases the alkalinity or carbonate hardness of the treated water.

In actual practice about 0.8 to 0.9 P.P.M of hydrated lime are required for each 1 P.P.M. of free CO_2.

1 HP = 0.746 K.W. = 550 foot lbs. Force per sec.(ft.lb.f.s.)

Unavoidable Loss in the Distribution System :

Several methods have been developed for estimating the same. Emil Keichiling says "A discharge of one drop per sec. From each joint, 5 drops each hydrant & stop valves & 3 drops for each service pipe including tap and unit cock represents as fair measurements of the average undiscovered leakage in a well constructed distribution system."

On this basis & with the assumption that on the average there are 504 pipe joints, 12 hydrants, 10 stop valves and 100 service pipes per mile of distribution pipe, leakage will amount to 2742 gallons per day per mile or to round up 2500 to 3000 GPD per mile.

APPRAISAL TECHNIQUES OF DWELLING UNITS

All India Institute of Hygiene & P.H. Calcutta has evolved a standard schedule for sanitary assessment of houses, in which credits

points are assigned - giving a score of 1000 for a house fulfilling all the requirements on the following pattern.

		Maximum Score
1.	General (including approach, type etc.,)	70
2.	Surrounding (openness)	80
130479184.	Cleanliness (a) outside (mosquito, fly breeding areas&drainage)	200
	b) Inside	20
4.	Living room (over crowding, Ventilation & lighting)	185
5.	Kitchen	80
6.	Water supply	130
7.	Garbage & Refuse	50
8.	Human Excreta	135
9.	Cooking Utensils	20
10.	Cattle shed	30

TOTAL SCORE –1000

1	66% and above	GRADE A	or	1
2	50 TO 65%	GRADE B	or	2
3	. 33 to 40%	GRADE C	or	3
4	32 or under	GRADE D	or	4

CHAPTER TWENTY

WATER TREATMENT

Sedimentation by gravity, or settling, is used to separate dusts, suspensions, and emulsions. Unfortunately, this method provides a low settling rate and fails to remove finely divided particles. It is mainly used for the partial separation of inhomogeneous mixtures.

Centrifugal sedimentation is the most effective method for the separation of dusts, suspensions, emulsions, and vapour (gas)-liquid systems. The required centrifugal force can be produced by rotating the inhomogeneous feedstock inside a stationary vessel (as in cyclones and hydrocyclones) or by rotating the impeller and the inhomogeneous feedstock together in a vessel (as in sedimentation centrifuges).

Filtration removes nearly all of the suspended particles from liquids and gases. There are two basic filtration mechanisms. In one, the filtered solids are stopped at the surface of the filtering medium and pile upon one another to form a cake of increasing thickness. This is known as cake filtration. In the other mechanism, the solids are trapped within the pores or body of the filtering medium. This is known as filter-medium (blocking or depth) filtration. The driving force is the difference in pressure upstream and downstream of the filter. It can be supplied by feeding the stock under pressure (as in pressure filters), by applying a vaccum downstream of the septum (as in vaccum filters), or by applying centrifugal force across the septum (as in filtration centrifuges). Filter media may be cotton, woolen, glass, or synthetic cloth, wire nets, porous materials, ceramics, cermets, and particulate materials (such as coal, sand, gravel, diatomaceous earth, etc.). Submicron particles are removed from gas suspensions by fibrous filters using glass paper as the filter medium.

Chemical methods of wastewater treatment include neutralization of acids and alkalis, conversion of ions to poorly soluble compounds, co-precipitation of in-organic substances, oxidation, reduction, electrolysis, hydrolysis, and catalytic oxidation. These methods are mainly used to deactivate and remove the impurities of inorganic compounds.

Physico-chemical methods include flotation, extraction, electrochemical and sorption methods, fractionation (distillation and rectification), reverse osmosis, and some others. The sorption methods include absorption, adsorption, and ion exchange. Except absorption, the methods listed here are used to remove finely dispersed, colloidal and soluble substances from wastewaters.

Absorption and adsorption are widely used to remove toxic gases, fumes and vapours from gases.

Physical methods include precipitation in electrostatic or magnetic fields, acoustic coagulation, evaporation, and some others. The electric field is widely used in electrostatic precipitators to remove solids and liquids from gases. The magnetic field is employed for the selective recovery of ferromagnetic particles, iron-carrying slimes, and other materials possessing magnetic properties from suspensions. Acoustic coagulation occurs when a gas carrying dust, fumes or mist is irradiated with ultrasound.

This causes the suspended particles to agglomerate, and speeds up their settling rate.

Biochemical methods are employed to purify sewage waters. They are based on the biochemical oxidation (biodegradation) of organic and some inorganic substances due to the life processes of microorganisms. Sewage water treatment uses two methods, one is aerobic digestion (decomposition or decay) in which there is a continuous inflow of atmospheric oxygen to the sewage water being treated. The other is anaerobic digestion (decomposition or decay) which proceeds in the absence of oxygen. Aerobic digestion is more versatile and more commonly used of the two. It provides for a maximum rate of biochemical oxidation and a maximum effectiveness in rendering the impurities inert and relatively pathogen free.

[Reference –source-Gannon drunkenly HAND BOOK, made by Shri Guha Thakutha]

CHAPTER TWENTY-ONE

Infrastructure-The law

The courts have limited time, and manpower to decide what law these agencies has to follow, when they are executing other's projects. The best suited arbitration laws are internationally complicated as the final verdict has to come from the court. The quasi judicial authority of the arbitrator has to be defined in the arbitration act.

The African nations traditionally look at India as a lead democratic country. But when it comes to administration, and corruption, they turn their heads to China, which is ready and willing to partner in their developmental needs.

The necessity to understand international law is imminent, for all. We can be on tour, for business, or for a job. But to deal with the laws of that nation where we step in, along with knowledge of the international laws is a must even for a small worker.

General Laws applicable in construction
1)   **Company Law:** Governs the board of directors, pay and benefits for   them, audit of company accounts   loss and profit over year responsibility   to share holders.
2)   **Shops & Establishment act:** This is applicable to the employees in terms of leave, terminal benefits, holidays. However an employee overseeing the functioning of the Company and finding faults in trade practices can directly write to Company law-board.
3)   **Contract lab our abolition act:** In this law the Principal employer need to give an undertaking that a contract is given to so and so Company, and that Company will employ on an average so many workers (type – category and no to be specified) for so many days. The Contractor goes to Dy. Commissioner of lab our files his application after paying (a) deposit for each lab our (b) License fee for each quarter. The Asst. Commissioner of Lab our will over see the (a) Attendance Register (b) leaves and paid holidays (c) Damages etc., caused by employee all registers are to be maintained each day and

quarterly reports to be submitted. In case of any disputes by Union or workers the Asst. Commissioner of labour is to preside and finalize the redressal.

4)

[BUILDING CONSTRUCTION]

5)   **ESI:** In a working 'manufacturing environment' the workers for whom license is taken are also to be covered under ESI by paying extra premium.

6)   **Provident fund:** Employer to contribute towards provident fund and family welfare fund (in case of accident) by paying about 8 ½% a (102pprox..) the employee to pay – 8 ½% for it and 4% extra for pension. This has to be done to almost all the workers covered under ESI, if not for the quantum for which lab our license is obtained.

7)   **Comprehensive workmen insurance:** In a single instance of tragedy (accident) how many workers are to be covered and the area of IN works (a) hazardous or (b) non hazardous. Locations to be specified.

8)   **Construction works insurance;** Against any mis-hap in which the works can be damaged – which are completed and billed either through the bank or to the client.

9)   Insurance for tools plant and machinery:
This is like third party insurance for automobiles. All Cars, motors bikes, Cranes, rollers, batching plants etc., need to be insurance. Generally those operating 'inside plant only' attract less premium.

[Cultural show, need for auditoriums.]

Remembering in 1987 I was in a general compartment of Konark [train] from Mumbai to travel to visakhapatnam,I found suffocated and breathless before I reached Pune.I was on top berth occupied by a couple of people, and with great difficulty got down at Pune,and took passenger trains to complete my long journey.

In years 1979-80 we used to travel by the side of bath rooms, in Konark express heading to Hyderabad from visakhapatnam Really it was better than losing lives of full train passengers in sleep, on a fateful night in November 2005.After all we came from a situation where a small kerosene lamp, directly burning in the air, was our source for living, be it for studies or for living till year 1965.

A similar startling experience lingering me since 1981 till date is air travel. Delhi was the worst in winter for takeoff and landing of flights. Flights are ordered to move to Agra due to poor visibility. There are [were] no parking space, hence the flights move to Jaipur, and after halt turn back to Delhi. The latest is congestion at; forget at railway stations, but also at the airports. Systems for airports have changed, but for the railways moving on ground other vehicles, winter fog and burning of wheat waste creates a very serious,hazrds.People and vehicles needs to take care.

I was living in a village house at Uran, New Mumbai in 1987, while working for a site work, as in charge. The heinous thing was the local 'Gabbar ' has dragged the earlier in charge[about a month before] to about a kilometer, for not attempting to declare his amount of commission on the works. That was about the construction job of a bottling company. The other is living condition.

The soak pit cum septic tank was in marshy land, and abutting the black cotton soil. The night soil was entering the latrine and the house at times of rains [on all rainy days].Life was looking so dirty, as no septic cleaners were available. After about 20 years in an apartment block in visakhapatnam,we find that we are neighbor's envy every 10 months, as our septic tank, releases night soil outside as it gets filled. The neighbor's always say we are the developed lot whom they envoy.

We the Engineering community building projects, went to places, where no proper Houses, no schools, we start from ashes, and when every thing for the industry is ready we are shunted out mercilessly, and un-cermoniously.The best is with no blame on construction quality. This is not always not possible. We have introduced septic latrines to a village named Chilamkur, in Cuddapah district, Andhra Pradesh, India, while building the then

Coromondel fertilizers cement plant. Now it is India cements plant.

Having worked with nearly a dozen companies in India, and associated with another dozen consultants, it was a difficult task to build 20 projects in 25 years in a senior, responsible, position. This is excluding overseas experience, and till 2005.

The horrifying experience was again in Uran, after an air travel from Bangalore, and taxi from Mumbai, I have to stay in a den called 'hotel', with family, for overnight. There was one entrance with a rolling shutter, no window any where. Not even a ventilator. I stayed in that concrete hole with no sleep in that night, with my wife and sons.

Another instance to quote was in year 1979, when i first traveled by 20-seater aircraft from Katmandu to Bharatpur in Nepal. It was horrifying; the way the flight flew over mountains surpasses all IATA regulations.

The best air experience was in Ethiopia, in years 2001 and 2002, without a control tower or a run- way at JIJIGA [50 kilometers away from Somali]. Please read Mr.M.J.Akbar experiences in Deccan chronicle, news paper dated 11-01-2006 and continued on 12 and 14]. You need to really thank the pilots. The other experience now a days is in Indian skies, a very young woman of 25 years is flying airbus between Chennai, and Mumbai, for Deccan air ways.

We used to get unlicensed explosives for our excavation works. There were no proper storage rooms as per regulations.

The other is the under developed attitude in soil investigation, and foundation engineering, for plants at many places including at Visakhapatnam, where marine soils are predominant. The unfortunate part is the companies don't delegate powers to small or middle managers about policy matters, but blame such persons to save the top, who take decisions in such matters, as a matter of right, in the companies. The company law board does not provide protection to these small employees.

In 1989 at Hazira, when the Bangladesh carpenters were threatening with strike, the good alternative for us ,from the company was to accept the Gandhian philosophy nonviolence. In another development in 1991,a train from Tatanagar reached to Bandamunda railway station, and found the track damaged. We the passengers were brought back to Tatanagar and asked to collect our refund for the tickets. There were some persons traveling to Madras for medical attendance, so we resorted for dharna and hunger strike for 4 hours, there by making the authorities to run the train in alternate route. Thus the Gandhian system worked. It was in 1995 we were told that a train from Bhuvaneswar will not leave till next day as the incoming train arrived late, again we went for Gandhian protest and succeeded in a few hours.

A beautiful experience was about the construction of Blast Furnace Receiving Station [electrical] at Tisco modernization in 1992.The design was on 30x23 cm brick pillars spanning at 12 meters. The iron beam on top buckled at center cracking the composite section of RCC slab. The designers tried to blame to constructors, latter understood the deficiency in design, and made modifications.

Another was the experience at the construction of quarters at Missile factory, at Bhanur, Medak, district, Andhra Pradesh, in year 1988.The huge balcony of about 1.5 meters for B type quarters was without column support and was over hanging. It buckled and failed; the designers sat at site to estimate the actual length of embed -dement for the beam bars from the cantilever in to the column.

Another experience is, with that of railway booking centre of Chennai].If an old or sick person is injured at these places, then are the departments, prepared to compensate them in court. The railway establishments are covered under labour act [rail unions] and hence the premises need to be verified by the inspector of factories. Similar is the case with many shopping malls in Chennai, whose finishing's beyond public utility. May be, they need a court order or a public interest litigation, petition in a court. The factories law is definitely applicable here.

Like the one said earlier, on a fateful day I was told in my apartment at Visakhapatnam, that all the parking lots, purchased and registered are cancelled [nationalized] by the district court. I was pained at the ignorance, rather than get angry at the arrogance. The court has ruled that the un-built and unapproved building area shall be handed over by the builder to the society, and directed to use such land for parking. No discussion or mention about the parking lots sold and registered by the builder was made in the court order. That is how precisely the understanding of common law is among the educated people world over.

I was the first to introduce septic latrine system in 1980,in a small village in cuddaph.

CHAPTER TWENTY-TWO

PILING

Total planning-for ports harbors and marine construction
Each of the following stages of project formulation, in urbanization is discussed below.
1. Site investigation to obtain, socio-economic survey.2.investigation to find technical feasibility, and cost studies.3.economic viability, parties for project finance.4.ultimate effect of the project, in terms of human development index.

Next step, understand the demographic needs, to arrive at the developmental requirements.

The population for which the project is intended. The present growth rate of population, and expected futuristic population in next 30 to 50 years.

The first stage in planning a site investigation comprises a desk study of the available information. The investigation should take into account details of the foundation design and permit decisions concerning the type of foundation.

Other specific aspects of the investigation may include matters such as the seismic risk o the area, aggressive soil conditions (especially in fills) and the possibility of aquifer pollution if piles are to penetrate contaminated ground.

The construction of piles today is also carried out for the most part by specialist contractors who exercise considerable skill in adapting methods of working to enable sound piles to be formed even in the most forbidding ground conditions, although it should still be borne in mind that every technique used can encounter difficulties and unforeseen problems at times. The skill of a particular contractor often depends on his particular experience and this is only gained over many years, perhaps using a range of methods.

[New large diameter suction boring rigs.These are for deep piles]

Housing- methodology

Since man first sought to establish secure dwellings and to cross streams and rivers where fluctuating water levels gave an element of uncertainty to his travel arrangements, the driving of robust stakes or piles in the ground has provided a means whereby the hazards of living could be reduced.

Various people in different parts of the world found it convenient to dwell by lake shores where food, water and easy transport were readily available to hand and where water levels remained reasonably constant. Evidence of piled settlements has been found on the borders of lakes in, for example, Switzerland, Italy, Scotland and Ireland.

It is believed that some of these settlements were in use about 4000 years go and they were sometimes of considerable size, as on the shores of Lake Geneva opposite Morgues. In another settlement at Roben hausen, it has been estimated that over 100 000 piles had been used.

Steam power had first been applied to the driving of piles in Britain by John Rennie in 1801 or1802 at the Bell Dock at the entrance to the London Docks.

The ram was hoisted by an 8-hp engine, constructed by Boulton and Watt, and this engine, or one similar to it, was used again at Hull docks in 1804. In 1843, Nasmyth produced a radical departure from this type of mchine, introducing a revolutionary type of hammer. In this new hammer the ram was attached to the lower end of the piston rod and the whole weight o the steam cylinder acted on the pileIt . achieved

first precast concrete driven piles appeared and although the originator is not known, the earliest drwings of precast concrete piles in the UK appear to have come from the Hennebique Company.

The use of steel I-beam piles originated in the United States before 1900, when fabricated sections were used for highway bridges in Nebraska, but after 1908 the Bethlehem Steel Co. produced rolled H- sections which quickly captured the market for this type of pile.

The use of steam-operated hammers continued throughout the first half of this century for driving piles of all types but declined in the post-war period after 1946, and in-place diesel-operated hammers became popular.

Steel H piles and tubular steel piles are used for limited applications. The H section pile causes minimal soil displacement and can stand up to fairly heavy driving conditions while the tubular pile finds application in many shoreline works where nboth direct load and bending capacity are often needed.

The design of piles and pile groups has advanced steadily in recent times with much of this work carried out by engineers who specialize particularly in foundation engineering. Effective stress methods are being developed for individual pile designs, while computer-based techniques for assessing the settlement behavior of large and complex pile groups are finding increasing application.

In modern piling practice site investigation has become an essential precursor to making sensible decisions both regarding pile design and the choice of appropriate construction methods, and is all the more important because of the wide range of methods and equipment now available. Financial savings made at the investigation stage are often followed by the unwelcome discovery of adverse conditions which can eventually impose an extra burden of cost far exceeding the earlier savings which may have been made.

Clearly it is essential that, with ever-increasing pile stresses which are consequent upon the strong incentives which exist to use materials economically; piles must be made both sound and durable.

honest endeavor eventually turn engineering concepts into dependable reality. It has ever been so.

Ravva off shore platforms [east coast].

The first phase was done by the ONGC, using Hindustan shipyard and ESSAR offshore. The RV-11,and 17,were connected on either side by the pipeline of different sizes. The S.Yanam Refinery became operational in 1994.Later Hyundai heavy Industries have done the second phase in 1996-97.The wells are under operation.

Further in to the sea is the Dhirubhai Ambani, gas field in operation approximately since 2007 onwards, with a refining unit in Yanam, near Kakinada.

Table 1:

QUANTITY	RIPPLE SAND WAVES		SANDBANKS
Wavelength (m)	10-2	102	103
Wave height (m)	10-2	10	30

Seabed Topography Charts

Navigational chats are probably the most common seabed topography charts. For safe navigation, it is sufficient if nautical charts give minimum depths. As the seabed is likely to change over time, it is expected that the accuracy of a chart will decrease with time. Apart from long-term influences, such as sea level rise and overall sedimentation o erosion, the combined effects of moving sandbanks, sand waves and mega-ripples form the major source of change in the least depths.

Echo Soundings

Nowadays, topographic data is mostly obtained using single-beam echo sounders, which are attached to ships sailing over the area to be charted. They measure the depth directly below the device. Recently, multi-beam echo sounders enable measurements not only directly under the ship, but also in a swath (strip) with a width of several times the water depth on either side of the ship.

Satellite Images

Remote sensing is adopted as an alternative way of obtaining data on the seabed topography. Satellite images are inexpensive and provide snapshots of the sea-surface. Techniques to translate satellite images into seabed topography are being developed at this moment,

The use of radar imagery for bathymetric mapping is under use with precise and accuracy. Because of the flow-dependence of the relationship between sea-surface patterns and seabed topography,

CHAPTER TWENTY-THREE

Pipeline & Cables

Hundreds of kilometers of pipelines can be found in, for instance, the North Sea. Pipeline protection takes up a large part of the total costs of developing a new oil or gas field [Li & Cheng, 1999]. These pipelines sometimes have to cross a sand wave field. The sand waves can form a thret if they migrate and expose the pipelines,

Free spans may develop, causing stresses due to gravity. Moreover, the pipelines can start vibrating, due to the turbulence generated under these free spans. The vibration also causes undesirable stresses, which may cause the pipeline to bend, break or buckle. Once exposed, a pipeline or cable can be damaged by ship anchors or fishing gear. The height and migration speed of sand waves are important design parameters for pipelines and cables. Mega-ripples are too small to create significant over-exertions.

Free spans may also be caused by changes in sand wave asymmetry, i.e. changes in the shape of the sand waves, irrespective of their migration, which may be caused by a change in the water movement across the sand wave field. Such a change in asymmetry may falsely be identified as migration, due to the large measuring errors. Hence, sand wave migration data in the literature are not always reliable.

The most straightforward solution is to lay the pipeline in a trench through the sand wave field. This solution is effective, but expensive. the main question is: what is the most efficient depth for a trench to place the pipeline in? This optimal depth depends on factors such as dredging costs, pipeline construction costs, monitoring costs and risk.

Furthermore, under certain conditions, the pipeline may have a 'burial potential' of its own. A pipeline laid on top of sand waves may be curved, when it will not sink as easily into the bed as it would in in the case of a flat bed. Moreover, the current velocity varies along a sand wave, making it harder to predict the burial behavior of the pipeline.

Knowledge concerning the behavior of the seabed (especially the migration rate of sand waves) and its interaction with a pipeline can help optimize the design to minimize the total costs. Since the pipeline follows the contours of the bed, we also need to understand the behavior of the entire profile of sand waves and to what extent the pipeline works itself into the seabed.

Survey Before, During and After Use;-Several bathymetric surveys are made in projects concerning pipelines. First, a reconnaissance survey is made. Next, the chosen route is measured more precisely. Before the pipeline is constructed, the route is surveyed once more to have the latest information about the condition of the seabed.

[Shore pulling arrangement for offshore pipe line laid in the shallow waters, with weight coated pipes. Testing the integrity of pipeline is done by using pigging. The travel of pig with ease is what gives the rate of flow across the pipeline for oil and water]

After the pipeline and its surroundings are checked. During the entire life span of the pipeline, this area is monitored on a yearly basis.

Knowledge about sand wave behavior can reduce the survey effort and the costs. It can improve the accuracy of the measurements and help their interpretation. **Channels,**

114

Navigational Routes & Sand Extraction

Access channels leading ports, such as Rotterdam harbor, have to be wide and deep enough for ships to pass safely, otherwise the channel needs to be dredged. Bathymetric information is provided to mariners so that they can navigate safely over the Netherlands' Continental Shelf. This is done by publishing nautical charts, depicting the least depths of the seabed together with information about wrecks and other obstructions.

[Sriniing a pipe line]

Both migration and seasonal variations in height or asymmetry of sand waves change the topography and possibly affect the minimum water depth. The channel plus its surrounding area has to be monitored in order to decide where and when sediment has to be extracted. Knowledge about the evolution of sand waves will make it possible to determine the depth to which they should be dredged if they form a hazard to navigation.

If a sand wave is lowered only marginally, the dredging costs per operation will be relatively low, but the frequency of these operations and the monitoring after wards are expected to be relatively high. Knowing the rate at which sand waves evolve will help determine the most efficient dredging strategy and monitoring interval to guarantee the least depths.

[Laying a offshore pipeline in back slushy-marshy lands]

Burial Of Objects

Objects lying on the seabed can get buried, due to the migration and/or growth of sand waves, after which time they lie dormant in the seabed, (Fig.4). Here one can think of objects such as ship wicks, mines and containers possibly containing hazardous materials. However, they might become exposed again, forming a direct hazard to the environment (for example, leakage of chemical waste). These objects can get stuck in fishing gear.

Knowledge about the time between burial and exposure (residence time), can optimize the monitoring strategy and thus educe the costs. there is little point in monitoring an area where the seabed hardly changes. Hereby, not only the horizontal displacement of bed forms plays a role in the residence time, but also the self-burial of the objects.

Even though sand waves are not directly visible to the naked eye, they pose a threat to a range of offshore activities. the combination of their timescale (years), length scales (hundreds of meters) and height (meters) makes them bed feature to be reckoned with.

The question asked by the institutions and the industries involved in offshore activities can be summarized as: "Under which conditions are sand waves dynamic (horizontal and vertical movement) and what are the typical spatial and temporal scales?. Furthermore, in a certain area, data are available over a period of

116

years, the evolution of sand waves in that area, data are available over a period of years, the However, the latter technique is based on data assimilation, so that the results are only valid for the location from which the data originates. A model based on physical principles describing the nonlinear dynamics is not yet available, but is expected in the near future.

Insight into sand wave migration is important to estimate the optimal monitoring frequency for navigation channels, pipelines and buried objects. Greater insight into the height evolution of sand waves is required to determine when and how much should be dredged if sand waves get too high and form a threat to navigation together with the optimal monitoring frequency.

[Pipe laying,jointing-welding,applying weight coat and burial]

A wise manager devotes his energy to considering when to listen and when to speak, when to act and when to observe. He should take decisions conscientiously and calmly. As soon as he has taken a decision, he should put doubt aside and unflinching steer the course he has set, even if the conditions he encounters are not the ones he expected.

CHAPTER TWENTY- FOUR

IRRIGATION AND HYDRAULICS

Normally rain gauge stations are established to measure the amount of rainfall in a given day. The gangue, the area are all standardized. These rain Ganges are like a network. Each rain gangue covers some area in square kilometers. Over the year the total daily/monthly rainfall in the area is established.

This gives over rainy, or flooding years. This also gives average rainfall per year in the area, and the drought years.

Normally, where crops are grown with rainwater, they require 50 cm. To 100 cm. Rainfall in 60 to 90 days. This is true in most dry crops, except rice, wheat, sugar cane etc., Rice requires about 2 M water in 120 days and sugar cane about 3 M water in 240 days. Since in most of the area the monsoon, rainfall is on an average 1 M, the remaining is supplemented by storage reservoirs, Dams etc.,

The rain fall in a given area partly goes in to ground water as per collation part of the rainwater gets evaporated. Only about 80% of rainfall is available as flowing water – called run off. Since, this flow is seasonal, it is not possible to store this water entirely, and hence rivers carry water to Oceans.

LIFT IRRIGATION SCHEMES / OR WATER SUPPLY SCHEMES:

Fundamentally water flows in Valleys, while habitation (or growth of towns) in on ridges. Hence, water is to be pumped up to a ridge point.

A sump well is made in the river; from this a suitable pump house is made at the nearby location. All necessary electrical and instrumentation are made near by.

A manifold is used to connect the water from a number of pumps to a pressure main. The pressure of water in these mains is 3 to 4 Kg. Cm^2. Prestressed concrete pipes or steel pipes with cement lining inside and tar felt coating out side are used to covey water, up to to a ridge, or a storage reservoir.

For irrigation this water is led into branch channels and field channels.

For drinking purposes this water is led to a clariflocculator for clearing organic matter. This water is then left to gravity filters and then required chlorination is done. Such water is pumped to overhead tanks, for drinking purposes in the city.

The pumps are generally vertical turbine or centrifugal pumps.

The pressure mains require air relief valves, reflex valves, etc., to avoid water hammer in the pressure main. Such a water hammer leads to sudden burst of pressure mains.

For flow through pipes in house connections and water supply the discharge

$Q = f \ lv^2/2gd$

Where f is a friction coefficient based on pipe material

L is length of pipe

V is velocity of water in pipe less than 1 kg/cm^2

G is gravitational force

D is dial of pipe

PUMP SELECTION PROCEDURE FOR HORIZONTAL SHAFT PUMPS.

Discharge through each pump is to be arrived after considering resistance losses.

Suction side:

Static suction lift=Pump axis level – LWL in sump well

Pump axis level=Platform level of pump + 0.3m to 0.5m

The dia of suction pipe is to be arrived for the discharge, keeping the velocity through the suction pipe within the permissible velocities given in enclosed Annexure-III.

a.   Entrance losses   = $KV^2/2_g$

    Where K is constant   =   0.5

    V   =   Velocity in suction pipe

    G =   Gravitational force I e., 32.2ft. /sec.^{2}

b.   Frictional Losses in
      Suction pipe & fittings

Approx. length of suction pipe=   25' to 30' approx. or the actual length.

Equivalent 900 bend   =   1 No.

Equivalent length of 1 No. Of Bell mouth or Foot valve with strainer.

Total equivalent length   = 25' + Equivalent length of 90° bend + Bell mouth or Foot valve.

Frictional losses in suction pipe can be obtained from the Williams and Hazen   Formula or form IS Standards.

Losses as per the Formula $= 4.69/D^{1.87} (Q/C)^{1.85} \times L$

Where   =   D = Dia of suction pipe in ft.

Q = Discharge through each suction pipe in Cusecs.

L=Equivalent length of suction pipe and fittings.

C+120 for MS or ERW pipe (Old or New)

Total suction lift including losses

= Static suction lift + Entrance losses = Frictional losses in Suction pipe & fittings.

Delivery side:

Static delivery head =   Delivery level at Cistern – Pump axis level.

The dia of delivery pipe is to be arrived for the discharge, keeping the velocity through the delivery pipe within the permissible limits given in enclosed Annexure – IV (A).

c.   Frictional losses in delivery pipe and fittings:

Approx. length of delivery pipe   =   25' approx. or as per Actual.

Equivalent length of Reflux valve =   1   No.Equivalent length of Sluice valve=   2 No.

Equivalent length of 90° bends   =   2 Nos.

Total equivalent length.

Total of head in delivery pipes and fittings can be obtained from the

Formula:

Losses as per the formula $= 4.69/D^{4.87} \times (Q/C)^{1.85} \times L$

Where   D =   Delivery pipe dia in ft.

Q=   Discharge through the delivery pipe in Cusecs.

L=   Equivalent length on delivery side.

C=   120 for MS or ERW pipe (New or Old)

Loss of head in the Manifold being small, it is ignored.

d.   Loss of head in pressure main: The dia of pressure main is to be arrived for the discharge keeping in view the maximum permissible velocities given in enclosed Annexure – IV (C). To arrive at the most economical dia. Of pressure main, alternates may be worked out changing the dia of pressure main, calculating HP, of each pump set and the cost, water rates etc.,

Equivalent length of pressure main length of 3 Nos.   = Length of pressure main

In ft. + equivalent of 90° bends (minimum)

Total length can be arrived at.

The loss of head in pressure main can be obtained from the Losses as per the Formula $= 4.69/D^{4.87} \times (Q/C)^{1.85} \times L$

Where   D   =   Dia. Of pressure main in ft.

                Q   =   Discharge in Cusecs.

                C   =   110 for pressure main of CC pipes or 130 for PVC pipes.

                L   =   Total length of pressure main ft.

Delivery and Pressure Mains for different discharges and diameters and types of pipes prepared the hydraulic Tables from the book written by William and Hazen. (Enclosed as Annexure-XV) These can be referred to, to arrive at the losses readily.

e.   Exit losses   = $V^2/2_g$

       Where   V   =   Velocity in pressure main and g =   32.2 ft./sec.^{2}

Total delivery head including losses = Static delivery head = losses in delivery pipe and fittings + loss of head in pressure main + Exit losses.

Total Pumping head of each pump:

= Total suction lift including losses + total Delivery head including losses +

½ dia of pressure main in ft.

Care may be taken that discharge occurs through the exit in the atmospheric conditions i.e., the pipe should never come under submerged condition.

The above total pumping head is used for finding the Ns (specific speed)

+ Add 10% on losses for ageing of pipes.

This total pumping head is adopted for the purpose of finding out HP of the pump

Specific speed at 1460 Rpm. (nom.)N= 3.65 X 1460 x Q/(h) ¾

Where Q = Discharge in Cum./Sec.

H=Total head including losses in Meters.

Note: For a Double suction pump A/2 may be taken instead of Q or specific speed may be divided by 2

Exact suitability of the pump can be obtained form the graphs as per IS 5120/1977 i.e., as per enclosed Annexure-VII. Alternatively if the graphs of American Hydraulic Institute are referred to the formula is

N3 = N Q/h

Where = N = Speed in rpm.

Q = Discharge in American gallons

H = Total head in ft.

The advantages are

i. Classification with respect to impeller design of the pump is done more precisely.

ii. Suction lift line on the Graph for the duty conditions (i.e., discharge and head) should be more than the total suction lift including losses. There must be a minimum difference of 0.50 meters.

If there is no such sate margin, the number of pumps may be increased. Even then if it is not within the limits. V.T. pumps have to be necessarily proposed.

HP Calculations

HP = 62.45 x Q x h/550 x n

Where Q =Discharge of each pump in Cusecs.

H = Total pumping head in feet.

n=Overall efficiency of the pump as given in the Enclosed Annexure – VI.

Cushioning is to be added as per enclosed Annexure – VI.

The HP is to be rounded off to the nearest range of HP.
Sizes of Sump-well, Pump house:

Based on the sizes of pump, motor and as per the dimensions given in the drawing (i.e., Annexure-I) the sizes of sump well, pump house can be arrived. Approximate sizes of pumps, motors are shown in the Annexure – VIII and IX, respectively.

Hydraulic Design of Approach channel and Intake:

For obtaining the maximum degree of streamline flow, flow to the intake or sump is trained right from the point of entry into the channel. Ideal conditions for the most satisfactory working of the pumps is, when it runs in still water. Since it is not possible to allow this condition, certain standards based on the model studies are adopted.

I)   Velocity in the approach channel should be restricted 1 to 1 ½ ft./sec.2
        0.3 to 0.45m/sec.

ii) Angle of approach should be kept within the range of 450 to 750 with reference to the   which the pumps are located on the axis normal to the flow of water in the intake. The angel of approach should vary directly with the quantity of flow.

iii) Portion of the intake or sump from the location of pump upto the length of longitudinal separators should have zero scope i.e., it should be perfectly leveled.

iv) To minimize the effect of interference when two adjacent pumps are working separators of specified length may be provided and raised to the height of max. Or at least normal level of water in the sump.

v)   Backing wall may be provided.

vi)   If trash rack is provided in a major LI Scheme it should be located at the entry of water into the separate compartments.

vii)   Enough margin in-depth may be kept for providing minimum submergence which depends on the cavitations, characteristic of the pump and which is furnished by the manufacturers.

Net positive suction Head calculations:

Available NPSH at LWL = pr. P.A

Where   Pr =   Atmospheric pressure – As shown in XII A to XII C

          P =   Vapor pressure as shown in XIIB.

          A =   Total suction lift including losses.

Max. Water level   =Platform level – 1.00 meter.

Available NPSH at Max. WL =Available NPSH at LWL + (Max. W.L – LWL)

The manufacturer of the pump should be asked to specify the NPSH required for the pump he is offering. There should be a clear margin of 0.6 meters between NPSH available at LWL and NPSH required. The NPSH available should be more by at least 0.6 meters than the NOPSH required. When this not the case, cavitation will occur during running of the pumps and so the particular pump is not suitable. We will have to go in only for such of the pumps whose NPSH required is less than NPSH available at LWL by the prescribed margin.

Selection of type impeller:

Whereas in centrifugal pumps, reading the graph itself will indicate the type of impeller which is required for the duty conditions as it is directly elated to the suction lift, it will not be the case in vertical turbine pumps as there is no question of suction lift. Hence it is absolutely necessary in selecting vertical turbine pump to heck the specific speed and the duty condition with reference to the configuration of the impeller design.

On American units, the specific speeds for different types of impellers are given in the following tables.

Type	Centrifugal Double-suction	Mixed flow Double-suction	Mixed flow propeller	Axial flow propeller.
n_s	1250	2200	6500	13500

Gpm	2400	2400	2400	2400
Head (ft.)	70	48	33	20
D_2 (in)	19	12	1750	2600
D_1/D_2	0.5	0.7	0.9	.0

These specific speeds may be compared with the total pumping head and it may be seen whether they are matching. If they do not change the speed of the prime mover according to the requirements.

CHAPTER TWENTY FIVE

Infrastructure-The executives from both sides

<u>Of the interaction between the government and the private agency.</u>

The executive can be termed as rouge, if he is envious of all he does, and wants more personal benefits. The need more service oriented Executives, in Public, Private and NGO sectors. The need of the hour for these countries is to evolve a code of ethics, for public and private servant as per existing laws, that the bribe giver and taker both are punishable under law.

[The bottom is extracts and comments on Rouge Executive, a book of washing- ton post by. 'Stanley A. Weiss', Business Executives for national security.]

O f course, the easy answer is greed. or hubris. Or any number of catchall explanations. But leaders seeking to avoid the same mistakes need more than generalities. To get ahead of the problem, the authors have reached decades into the past, seeking to untangle the complex Knot of personal motivations and systemic flaws underlying the Widespread pattern of deceit revealed over the past several years. The book reads like the voice-over from a hard-hitting documentary--often dry but unforgiving. As the camera pans, we examine macro trends, such as the rise of 24-hour news and the persistent drumbeat for quarterly earnings, and their micro results: imperious CEOs and their unscrupulous accountants. It's impossible to ignore the message. Patching up accounting loopholes with fixes such as Sarbanes-Oxley addresses only the symptoms. The deeper causes—the slippery whys embedded in our systems and ourselves--need closer examination if we want to avoid marching down the same paper trails again.

Back story, Enron, WorldCom, and Andersen--many of the book's villains will be familiar. But Sayles, a management Prof. at Columbia's B-school, and Smith, an anthropologist who chronicled Andersen's collapse, also sort through the rubble of more than 100 lesser-known corporate scandals to reveal broader trends.

What we liked is that this book digs deep, looking beyond the fallout to the root causes behind the mess. More important, the authors explore the fundamental weaknesses of people and organizations with an eye toward how they can be mended. Even with all the wrongdoing, the book's tone is pretty optimistic. What we didn't, At times the writing is flat and disjointed, leading to flashbacks of sleepy lectures in econ 101.

Some of the material seems to be from anonymous sources ("an executive at a company says"), leaving readers with vague assertions and not much hard evidence. What to say to sound like you've read it we've spent years cataloging C-level chicanery.

It's time to stop compiling the offenses and start using what we know to build solutions, not just more rules, into the system.

[A LIFT IRRIGATION SCHEME FOR IRRIGATION AND DRINKING WATER]

Infrastructure Management-

The word management has become synonymous to differentiate the owner, and the investors, and the people who implement the agreed programme of the agency or the company. The owners were represented by their 'chela's or agents in the nineteenth century. The concept of management for mixing various technologies has become imperative in modern day living. Technologies cannot live in isolation, and needs to be integrated one with another.

The use of robotics, artificial intelligence, censors, are imperative in mechanized world of infrastructure construction. The creation of each system is first drawn from a reflective animal and latter optimized with human reflections. The examples are, a jaw crusher reflecting human jaws, excavator reflecting a human hand. Compressors and air filters are replicated from human lungs. Ball mills and pulverizing systems are made to function like human digestive system. *The cylinder and piston which generate energy are derived from male female mating.*

Coming to human endeavor of modern infrastructure, today in 2022, bricks and tiles are as old as 2600 years. Mud and bamboo along with wood continue to occupy the modern age constructions. Even steel as an alloy is as old as 1000 years. Palmera leaves still are predominately used for village constructions in Asia and Africa. The old age construction of wooden piles in sea water has been taken over with deep foundations of concrete and steel.

In spite of all these the nature is playing its usual role, the role in which many riverside, sea side civilizations have vanished. We say our capabilities are increasing after each calamity. This is how the survival of the fittest has happened in the history.

Human knowledge in waste and water management is as old as 2500 years. People graduate with same text books, but differ in conceptualization and implementation of the infrastructure projects. The end results can be audited by anyone. Man has come a long way, still the ancient form of housing are safe against many natural calamities, and disasters. Bamboo net with mud on either side forms walls, and roofs made in the same way are most safe. This construction is as old as Greek civilization, and is still prevalent in Asia and Africa.

The bamboo fibers are netted to make the mat. The mud has the characteristic of being workable when water is added and shrinks when water is dried out. Thus the same mud is used for joining of broken bones. It was also used to kill a person who was found guilty. How ever the killing was slow. The sloped roofs with tiles instead of leaves started in heavy rainy areas over 800 years ago.

129

The usage of long planks and or thin stone slabs in sloped roofs can be seen in kerala state of India. The forms of structures that withstands many natural disasters are long chimneys [or towers], which are designed as self supporting structures.Charminar, Kutub-minar, etc

The significance is the shallow and vanishing Gulf waters, along with many marine and biological lives visually available on land. The other is the salinity of land still predominant with biological growth of natural tress not in the vicinity.

Remember a common sense could have stopped The Island express train [Sri Lanka] from running further, after the first tsunami wave on December 26 th 2004.Similarly commonsense on any one's part would have seen that the Rapalee - Secendrabad passenger [India] is delayed by 5 to six hours in very intense rains. These are all fateful infrastructural deficiencies, at the planning stage, later to be blamed on some one else.

The urbanization as a means of development and as a social tool is recognized, about 2000 years back, when Greek and Roman cities were built. The elite used mahanjadaro, and other ways of life, at those times. At times we teach and praise these cultures as a modern way of living and fail to augment our systems to meet those set standards over a period of time.

SCHOOLS;-The least and last is the premises for schools. Why the departments cannot choose Charted Engineers to set, and verify national housing standards. Even the SIDBI [small industries development bank]is not shy in taking expertise from engineers of repute. Need is first aid doctors, well-ventilated number of toilets for Girls and women, with sufficient water to meet the Menstrual needs of the young and as well perimenopause needs of older women.

A hydrant system at insurance towers, public offices, crowded bazaars, and secretariats has to be planned, at lest now. I shall continue the episodes as i proceed further. Fire hydrant systems are never thought of in any public places, and conveniently ignored at industrial establishments, and entertainment places, railway stations, bus complexes, etc.

A poorly standardized school in an urban area [photo]A school with out a playground, a gymnastic room, an auditorium are most common. Given a chance, many of the schools do not have minimum laboratories. Toilets for the boys, Girls, and the disabled are a must. Depending on school strength, the number in each floor has to be designed. A fire hydrant system, a way to come out during damage, accidents, and fire are to be established.
A competent engineering association has to certify the availability of such ,on yearly basis.

A government school in a city without a playground.

I know the orphaned families, the fathers and mothers because of the reckless thinking of their wards staying abroad. The limited personal resources available, in terms of money, job, and time are making the human bondage an impossible task in the lives of many. The job market is hapless that studying a bachelor's degree in engineering, by hook or crook is also not fetching a job.

Thus, the deficiency in The education system, lack of employment opportunities, and the caste based reservations in the existing system is making many middle class youngsters to migrate for foreign countries, preferably developed countries.

The other story is many aggressive agents for gulf job market. who are collecting as good as 100,000 Rs,for a job of labour with a monthly earnings of say 60 Kuwaiti dinars. The cost of food is around 30 Kuwaiti dinars. Exploitation everywhere, with no checks and balances contemplated in the constitution? Every authority supervising feels they are above law.

Roads

Ration-space on roads, a necessity of future

In modern day there was forced ration in times of war. Now the days are ahead for developing countries whose populations are on raise to adopt for ration systems on roads. Pedestrians have a right to walk.

The remaining place on urban roads can be rationed depending on the necessity, as it is difficult for us to make 8 lane [4 lanes in each direction] roads. The necessity to peg down automotives also arises. The e vehicles may increase then clusters for rest and charging needs to be planned.

Mass transport system comprising of, Railways, surface transport Buses, and future choppers for company use are to be planned for anticipated increases in the urban population and other modern developmental needs.

[A multi story buildings in use]

Infrastructure-Transport system

Transmission of energy, as well oil and gas distribution, along with creation of jobs, and planning mass urban production, and distribution are the basics for infrastructure planning and implementation in urban areas. Agriculture policy, land use, pricing of agriculture has to remain with governments till as late as 2030.The agriculture produce storage and marketing has to remain with governments. So much so said why the westerns cannot privatize their mass transport systems, and health care, apart from education, shall be an eye opener for the Afro-Asian nations.

CONTRACTS-ARBITRATION

"Construction contract" is an "agreement" with a person or a company by the owner. However, "agreement" is only one of the essential ingredients of an enforceable contract, the others being consideration and an intention to create legal relations.

A letter of intent for instance may be acted upon and constitute an agreement to carry out certain works, but may not give rise to a contract.

The Indian Contract Act came into force on 1 September 1872. It was enacted mainly with a view to ensure reasonable fulfillment of expectation created by the promises of the parties and also enforcement of obligations prescribed by an agreement between the parties.

The Third Law commission of British India formed in 1861 had presented the report on contract law for India as Draft Contract Law (1866). The Draft Law was enacted as The Act 9 of 1872 on 25 April 1872 and the Indian Contract Act, 1872 came into force with effect from 1 September 1872.

Construction Contracts Act

This Act deals with payments due under construction contracts. Find out what your rights and responsibilities are and how to use the adjudication process under the Act if you have a dispute about payments.

Purpose of the Act

The Construction Contracts Act 2002 came into force on 1 April 2003. The purpose of the Act is to provide a process for deciding what payments have to be made under a construction contract and when they are due. When you enter into a contract with a builder to build your house, you have to comply with some sections of the Act.

The Act also provides quick and simple procedures to resolve disputes about money due under construction contracts.

What sort of work is covered by the Act?

The definition of construction work in the Act is wide-ranging and covers most of the work which might be done in building, altering or maintaining a house such as:

The construction, installation, alteration, repair, restoration, maintenance, extension, demolition, removal, or dismantling of any building or other structure.

Putting in roadways, pipelines, water mains, sewers, electricity, water, gas, or telephone reticulation. Work done on any heating, lighting, air conditioning, ventilation, power supply, drainage, sanitation, water supply or fire protection, security, and communications systems. Any cleaning work carried out as part of the construction work. Preparation for construction work such as site clearance, earth-moving, excavation, laying foundations; scaffolding. Site restoration and landscaping, painting and decorating.

Payment schedules in contracts

The Act says when you enter into a contract with a builder (or other type of contractor), you can negotiate a payment schedule with them to cover:

The number of payments.

The amount of each payment.

The date when each one is due.

We can't have a clause in your contract that says you won't pay the builder until you get paid by someone else – this is known as a 'pay if paid' clause and it has no legal effect.

If we are building the house for someone else to live in, say as a 'spec' [define] house or a rental, this would be a commercial construction contract (which means "a contract for carrying out construction work in which none of the parties is a residential occupier of the premises that are the subject of the contract" in section 5 of the Act).

The default payment provisions in the Act, which apply if you don't have a payment schedule in your contract with the builder (or contractor), are:

At the end of each payment period (usually at the end of the month) the contractor can serve you with a payment claim. This must be writing.

Once received , we can either pay the claimed amount before the due date (20 working days after service of the payment claim) or provide a payment schedule to the contractor.

What happens if you fail to pay?

If we don't agree with the amount being claimed by the builder/contractor, we can give them a written payment schedule stating the amount should pay, need to say exactly:How it is calculated.

If we simply don't pay and don't give the builder/contractor a written payment schedule, it becomes a debt that they can recover from you in the Courts, along with legal costs.

If we do provide a written payment schedule, but don't pay the amount you've proposed on the date due, they can recover your proposed amount, along with costs, in Court.

The builder/contractor can only suspend work if this is agreed in the contract, or if it's a commercial construction contract, i.e. if they have built the house with the intent of selling it or renting it.

If we don't pay because we are disputing the scheduled amount, we can use the adjudication procedure introduced by the Act.

Adjudication

The Act provides a fast-track adjudication process for disputes. The only disputes that can be referred to adjudication are those which relate to payments under the contract and any disputes the contractor have about the rights and obligations of the parties under the contract.

The adjudicator can make a decision within a very tight timeframe and the decision is binding on both the owner and the contractor. It is also enforceable as a Court judgment giving access to the normal range of enforcement procedures. The contractor can also ask the adjudicator for a charging order over the work.

Nothing in the Act, however, prevents both parties from first submitting a dispute to another form of resolution, such as the

Courts, arbitration or mediation.

How to start an adjudication

Whoever starts the adjudication process must serve a written notice of adjudication on the other person giving details about the dispute and the names and addresses of both parties for service of the legal documents.

The adjudication processes

The steps in the adjudication process are:An adjudicator is chosen ,by the owner and the builder/contractor agree on one.

Three is Indian Arbitration Act amended from time to time,and contract labour abolition act ,which primarly concerns all construction in India.

For an Engineer what are the basic needs in communication and records for claiming an extension of time or a change in price.Some times extra items needs to be done how best the Engineer as the representative of the company address all.

[1]Daily Labour report-Weather the contractual claim of completion by barcharts has sufficient work force or not[2]The material at site,this includes the material needed for fixing in works and the material as form work,or reinforcement.[3].Daily progress report.When the program of client is broken for each month are the workers have sufficient skills to deliver to the speed and needs of the client.

The client also assess the speed of work and try to reason out payment history and the availability of funds

[4].The cash flow -is the needs of site in terms of money are met or not.

The engineer needs staff who understands

[1] The needs for Bar bending schedules preparation as per the IS-456.

[2] He needs subordinates who can measure and report daily labor output. That means he needs to understand the estimation of quantities. Need to know how estimation and costing are done.

Knowledge of Excel and his estimation book. Also IS 1200 ways of measurement.

Also, the standard schedules of Rate adopted by states for each circle. The data book of the state governments. A similar system is adopted by the CPWD for data and SSR.This practice is in each country.   Fortunately, the governments are uploading these statistics in their websites for use. They are available free of charge for down loading including the Code of practices for Buildings and for Reinforcement and steel etc.

The total system of Preliminary conditions, Technical conditions, special conditions, the EMD,the Security Deposit and the Mile stones, and the work force etc form part of a contract.

It is the Duty of client to hand over unhindered approach to the sites planned for work.

For a graduate Engineer to be a successful Resident Engineer/Project Manager is to understand the above and have the knowledge of materials, the testing needs and the design criteria.

Learing and teaching are a continuous process in the construction sector.

The position of the earth during its revolution around the sun deter-mines the amount of insulation or the incoming solar energy, the two hemispheres receive. This results in seasons. Thus the summer, or win-ter Solstice, and northern spring or southern spring Equinox, are be-cause of the earth's movement around the sun.There is a change in the orientation of Earth's axis with respect to fixed points in the celestial sky. This is called precession of the Equinoxes.

As a result, the axis traces a conical path I the sky. This conical path is due to the gravitational forces of Sun and Moon, acting on the earth's axis and its resistance to forces. It takes 26000 years for the axis to complete tracing the conical path once.

This, along with the large 'aspect ratio' (the ratio of the horizontal scale to the vertical scale) of the fluid stretched out on its spherical surface, has profound implications for the dynamics of the ocean. For large-scale motions, the force due to the pressure gradient tends to balance the **Carioles force – this is called geostrophy. A consequence is that currents tend to flow along isobars, rather than across them; in the** Northern hemisphere, a geostrophic current flow with high pressure on its right.

Another consequence of **the Cariole's effect is that the stress exerted by the wind on the surface of the sea does not set up a downwind current**; the current at the surface flows at an angle to the wind, and the transport in the surface boundary layer, called the Ekman layer, is to the right of the wind in the northern hemisphere.

Further the rotation of earth on its axis, and around the sun makes it to come back to a geo-position of any place on the earth to the same cos-mic position, takes approximately 60 years. This was how during salivahana period, 60 rotational years names were attributed for calendars.

As a result, the axis traces a conical path I the sky. This conical path is due to the gravitational forces of Sun and Moon, acting on the earth's axis and its resistance to forces. It takes 26000 years for the axis to complete tracing the conical path once.   For statistical purposes, we have earth diameter at equator is 12,756.8 kilometers. Through the poles the diameter is 12713.8 kilometers. Equatorial circumference is 40,077 & polar circumference is 40,000. Area of sea floor is 70.78%, and area of land is 29.22% or 149 million square kilometers. Area of Asia-is 44 million square kilometers, or 29.5%, area Africa-20%..

If some Civil Engineers who has learned Astronomy, Railways, Ports, Highways in their graduation they will understand about the pole star and the observations around it.

Tides and variations; Tides are caused by the gravitational pull of the moon and sun on the water which covers the Earth. Because the posi-tions of all three bodies can be easily projected, scientists can predict the tidal patterns for an area and create a tide chart.

Tides are not uniform around the Indian seas.

Even the south and northern seas around gulf of combay[Gujarat] vary very much.

 A tide chart also takes the topography and history of the region into account, and is a very specific document designed to be used in a small area. Generally, there are four tides a day: two high tides, and two low tides. The variation in height between the tides depends on location, season, and astronomical phenomena; sometimes the variation is only a few feet, while on other occasions it has been recorded to be as much as 50 feet (15 meters). In most cases, a tide chart will also include a graph to help readers visualize the differences between the tides.

Usually, a tide chart will be published for a general region with a list of corrections for specific areas in the front. For example, you may need to add or subtract minutes to the time to get an accurate estimate of when a high or low tide is going occur. In addition, there may be variances in height, which are also listed in the regional corrections. When reading a tide chart for a region, first check to see if you need to make corrections, and then open it to the relevant day.

Make any adjustments needed for an accurate reading, and plan your day accordingly. The daily tides have two high tides and two low tides. The fortnightly tides increase or decrease occurs depending on season. How ever the full moon days or immediately after offers high tide. The tides occurring in December reflects the movement of water from northern hemi sphere to southern hamispheare.

That is due to the effect of sun's gravity effects on each region of earth based on its rotation, round the sun as well around it self. Thus the tides in northern hemi-sphere starts dropping [rather sea levels dropping by about 2 meters] by about 25 th of December, and starts increasing by around 25 th May onwards. Any ocean disturbance at these times near equator sends back huge volume of water which is crossing equator at that time.

Thus the lowest of low tides will occur for us in Mid march, of each year, and any ocean disturbance shall not effect India.

Seven Minutes, written by Irving Wallace, he quotes the movement of moon and its implications on human behavior. More so the behavior is not expected on a full moon full tide day. All men under stress behave in the same fashion,the Doctors in UK found that doing a heart surgery near to full moon will be more successful, and the reversal on the black moon day.[These are some quotes-belive or donot belive]

Remember the world is a factory in making, but consistent to nature. If one feels this is absurd the following examples, illustrates this.

The earlier civilizations were concentrated at few places. The invention of fire and use of milk has made them more dependent on animals. Later the animal with water availability has made agriculture another support for life system. Family structure made man committed to provide food to such in family, who are physically sick. Later the village and community concept led to commerce. This has led for village industries of pottery making, and making tools for cultivation.

The buildings shall have an inbuilt intelligence to alarm theft and burglary. The foundations and roof tops shall have robotic systems to detect unforeseen flying and landing of helicopters, or planes.

The sensors, and transponders, will warn by alarm when water levels increase in foundations, or the soil underneath gets loose, due to earth quakes. Thus, for each of the above factors, an alarm rings in living rooms and bed rooms. We need to provide springs with elastic beams in the footings to with stand the S and P waves arising out of the new formation of earth quakes.

[Nairobi-The financial district of Nairobi, Kenya.]
We have knowledge without boundaries.
Intellectual property rights. All in one.
1] Hands on experience is a must in operating survey equipment, for surveyors, supervisors, and engineers.
[2] Real time project cycle experience, in day to day work of each work is needed as an apprentice, for all in to construction.
[3] Real time projects cycle, in construction is a must. Each project is unique in its nature.
[4] Assessing physical, quality of materials and testing them in laboratory at intervals is important.
[5] A conceptualized schedule, weekly and monthly needs to be arrived and monitored regularly at each stage of hierarchy

[Ethiopia [East Africa], Addis Ababa]

. In planning for a building or a town ship the following items needs to be taken care of. The trainer has to be taught the imperative need to speak about
[a] the foundations
[b] the layout
[c] Checking the soils, and rock
[d]The tests for aggregate, sand water cement as per the concerned code of practice and [e]the layout and survey.

Always remember the principles of national building codes, and important CODES concerning the construction.Estimation,and minimum design needs are a necessity in using equipment in erection and avoiding accidents.

The next book will be
"Instant flooding Hazards for Cities"